Creep – Fatigue Models of Composites and Nanocomposites

LEO RAZDOLSKY
President, LR Structural Engineering, Inc.
Lincolnshire, Illinois, USA

CRC Press
Taylor & Francis Group
Boca Raton London New York

CRC Press is an imprint of the
Taylor & Francis Group, an **informa** business
A SCIENCE PUBLISHERS BOOK

First edition published 2023
by CRC Press
6000 Broken Sound Parkway NW, Suite 300, Boca Raton, FL 33487-2742

and by CRC Press
4 Park Square, Milton Park, Abingdon, Oxon, OX14 4RN

CRC Press is an imprint of Taylor & Francis Group, LLC

Library of Congress Cataloging-in-Publication Data (applied for)

ISBN: 978-1-032-21301-9 (hbk)
ISBN: 978-1-032-21302-6 (pbk)
ISBN: 978-1-003-26772-0 (ebk)

DOI: 10.1201/9781003267720

Typeset in Times New Roman
by Radiant Productions

Preface

Difficulties in determining the resources of engineering objects are directly related to the complexity of the processes that occur in structural materials under the existing operating conditions. Understanding the laws of these processes will allow building a reliable mathematical model that will contain specific parameters of the stress-strain state (SSS) that meet the working conditions of the object, and which, ultimately, can become the theoretical basis for creating methods and algorithms for assessing the resources of objects in accordance with the individual histories of their use. Since the processes of damage accumulation are closely related to the kinetics of the SSS, the accuracy of the calculated estimates of the strength and service life of the structure elements will depend on the degree of certainty with which the ratios determining the mechanics of defective materials (MDM) represent danger zones of deformation in structural elements at predetermined operating conditions. Viscoelastic deformation parameters such as length and type of trajectory, type of stress state, history of its change, and others significantly affect the rate of damage accumulation. Thus, the main goal of research in the field of mechanics of a deformable solid is rather not to clarify the various formulations necessary for determining macroscopic deformations from a given loading history, but to seek to understand the basic laws of phenomena that prepare the ultimate state of the material and structure until it fails. This manuscript represents a logical extension (in terms of applied methods of mathematics used in this book) of the author's previous book "Phenomenological Creep Models of Composites and Nanocomposites" that was published by CRC (Taylor and Francis) Company back in 2019. The subject matter of the manuscript offered here is significantly different, since the phenomenon of fatigue-creep behavior of composites and nanocomposites at high temperatures is considered. In recent years, the use of composites and nanocomposites in various aerospace, automotive, marine, and civil engineering applications has been constantly increasing. Without exaggeration, it can be said that it is among the most complex and urgent problems of the mechanics of deformable solids, since phenomenological models in this case are much more complex, since they have to take into account an additional number of factors arising during cyclic loading. They are primarily associated with the development of cumulative damage. Currently, the problem of structural strength under cyclic loading is considered in a much broader sense. This is due to the development of new industries with modern technology, such as aircraft, power engineering, aviation and rocket technologies. The cyclic loading significantly reduces the creep-fatigue resistance in the entire frequency range, and it became obvious that such traditional characteristics of strength as endurance limit, static creep limits and long-term static strength can no longer suffice the design criteria for reliable performance. As a result, new directions appeared in the areas of high-temperature strength-cyclical creep and long-term cyclic strength. These circumstances led to the creation of new methods and means for determining the resistance of composite and nanocomposite materials and a continuum of damage development under cyclic loading; to the creation of appropriate physical models. Particularly relevant is the intensification of creep by high-frequency cyclic loading in composite materials, which usually occurs at high temperatures. Most of the known studies in the field of cyclic creep are experimental, and the direct use of the number of cycles to define the damage model cannot escape the empirical relation that predicts fatigue life level.

This chapter presents new phenomenological cyclic creep-fatigue models for describing the fatigue life and behavior of time-dependent composites and nanocomposites. Certain criteria imposed on selecting the creep functions have the potential to describe a wide range of material behaviors. Unlike metals, composite materials are substantially heterogeneous and anisotropic. This author presents the new phenomenological cyclic creep-fatigue models describing the fatigue life and behavior of time-dependent composites and nanocomposites. Damage does not accumulate in a localized form, and failure does not always occur because of the propagation of a single macroscopic crack. Due to the high specific stiffness and strength of composite materials, they are often used for weight-critical structural applications. However, existing imperfections in the methods for evaluating the strength of these materials often leads to the fact that large safety factors are required in structural calculations and designing. Although the technology of composite materials is developing rapidly, their use in real structures is hampered by the lack of sufficient reliable experimental fatigue data, which contains the main operational parameters. The analysis of the stress state under variable high temperature loads presented below requires taking into account the anisotropy of the averaged elastic properties of the composite and nanocomposite materials. The book consists of five chapters. The first chapter considers issues related to the formulation of the creep-fatigue problem in general on the basis of the classical Volterra integral equation of the second kind in the case of nonlinear creep. Of course, this integral equation is supplemented with a periodic function reflecting the fatigue process of a composite or nanocomposite. In addition, the kinematic differential equation of the continuous damage function has been added. This chapter also provides the required parameters that are obtained from simple tensile (compression) experiments on a composite sample. In order to "test" the proposed model, an example of a creep-fatigue calculation of a simple mechanical element of the Kelvin – Voigt model is given.

The second chapter is devoted to the application of the cumulative damage function to the solution of the problem of creep-fatigue of composites or nanocomposites under the action of high temperatures. Without loss of generality of the problem posed, some restrictions are introduced on the function of accumulative damage and the integrand of the Voltaire equation of the second kind. In the case of nanocomposites, the functions of cumulative damage consist of two parts, reflecting the technological process of creation (nucleation and growth of clusters) under the action of high temperatures.

In the third chapter, various phenomenological models of the maximum lifetime of a composite or nanocomposite are considered. At the same time, it is noted that some parameters of the system (such as the volumetric percentage of any filler or the temperature dependence of the viscosity coefficient significantly affect the life of the composite. In the fourth chapter parameters and variables (dependent and independent) in dimensionless form are used to obtain fatigue curves analytically. It should also be noted that, in addition to the usual dependence ($a_0 – N_f$), it is also possible to obtain other dependences of the cumulative damageability (ω - N_f). In this case, the limiting state of the system is $f(\omega) = 1$ @ $\Theta = \Theta_{max}$ or the usual dependence of the fatigue curve ($a_0 – N_f$) of composites and nanocomposites. Thus, the creep-fatigue problem is solved in a deterministic formulation. In structural engineering practice an empirical so-called safety factor is usually introduced. Chapter five contains various methods of the applied theory of probability applied to evaluate this coefficient qualitatively and quantitatively. It appears inevitable that the structural engineering community, as well as many other engineering communities that are ultimately responsible for life safety issues, will eventually incorporate probabilistic analyses methods to some degree. Probabilistic analyses methods, unlike traditional deterministic methods, provide the means to quantify the inherent risk of a structural design and to quantify the sensitivities of the most important parts of the design in the overall reliability of the structural system as a whole. The degree to which these methods are successfully applied depends on addressing the issues and concerns discussed in this book. The importance of a probabilistic approach in the issues of the cyclic creep-fatigue process and the long-term behavior with a specified reliability is addressed in NASA reports and other technical publications. One of the prime issues for these structural components is assured long-term behavior with a specified fatigue life. The present book addresses issues pertaining to the probabilistic creep-fatigue life of composites under combined thermal/mechanical cyclic loads through some typical examples. The focus of ongoing research has been to develop advanced integrated computational methods and related computer codes to perform a complete probabilistic assessment of composite structures. These methods account for uncertainties in all the constituent properties, fabrication process variables, and loads to predict probabilistic micro, ply, laminate, and structural responses. The results are supported by step-by-step practical design examples and should be useful for practicing structural engineers, code developers as well as researchers and university faculty personnel.

It should also be noted that some restrictions are introduced in this book (not violating the generality of the issue under consideration), namely:

1. All differential and integral equations, as well as independent, dependent variables and their parameters are presented in a dimensionless form. At the same time, in accordance with the theory of dimensionlessness, it is possible to achieve the minimum number of parameters and independent and dependent variables.
2. Solutions of all integral and differential equations describing the behavior of composite and nanocomposite materials under the action of static, dynamic and high temperature loads are presented in the form of numerical integration of the corresponding equations. This computer output is presented in three parts: systems of differential equations, explicit inputs, and implicit inputs. Due to the fact that these output datasets differ only in the value of one load parameter, in this case the complete solution (all three parts) is given for the smallest load value, and in all other cases the system of differential equations and the desired expression of a particular function (for example, the dependence of the oscillation amplitude on the variable temperature) are given.
3. The proof of the mathematical method for solving integra-differential equations is purely mathematical (Existence and Uniqueness Theorem for Volterra Integral Equations of second rang) and not included in this book. However, the simple result of this theorem in this particular case is as follows: in order to substitute the Volterra Integral Equations with corresponding Differential Equation the Initial Condition must be added.

Contents

Nomenclature

A	constant
A_i	unknown parameter
[a;b]	interval
E	modulus of elasticity
C; C_1	specific creep
D	diffusion rate, D is a constant,
Q	is an activation energy
J	compliance function (often also called the creep function)
K	stress memory function
S d	esign load
R	structural resistance
t and t_1	time $t > t_1$
T	temperature
$_t$	retardation time
$_{tr}$	relaxation time
η	viscosity
α	material property parameter
ε	strain
σ	stress
p	frequency
N_f	number of cycles to failure
ω	continuum damage function
φ	volumetric concentration
F	force

$$\theta = \frac{E_a}{RT_*^2}[T - T_*] - \text{ Dimensionless Temperature}$$

$$\beta_1 = \frac{RT_*}{E_a}\ ;\ k = \frac{E_m}{E_f};\quad T_* \text{ - Base Temperature } [^0K]$$

Time: $t = \dfrac{h^2}{a}\tau$ [sec]

Temperature: $T = \frac{RT_*^2}{E}\theta + T_*$ [K], where $T_* = 600°K$ is the base line temperature

k The thermal conductivity that has the dimensions W/m*K or J/m*s*K
T Temperature
d thickness in the direction of heat flow.
ρ is the air density
c_p is its specific heat capacity at constant pressure
K is the number of collisions per second resulting in a reaction A_t is the total number of collisions
E activation energy
R is the ideal gas constant
P Heat loss due to thermal radiation
e Emissivity factor
σ Stefan-Boltzmann constant (σ=5.6703(10^{-8}) watt/m^2K^4);
T_o ambient temperature
c_p average specific heat at constant pressure
t time
$\vec{v}(u; v; w)$ velocity vector
D Diffusion coefficient [m^2/sec]
p is the pressure
ν kinematic viscosity; $\nu = \mu/\rho$
θ dimensionless temperature
"τ" dimensionless time
"h" height of the specimen [m]

"a" thermal diffusivity [m^2/sec]

Time: $t = \frac{h^2}{a}\tau$ [sec]

Temperature: $T = \frac{RT_*^2}{E}\theta + T_*$ [K], where $T_* = 600°K$ is the base line temperature

Coordinates: $\bar{x} = x/h$ and $\bar{z} = z/h$ - "x" and "z" – dimensionless coordinates.

Velocities: $\bar{u} = \frac{\nu}{h}u$ [m/sec] and $\bar{w} = \frac{\nu}{h}w$ [m/sec] - horizontal and vertical components of velocity respectively ; ν – kinematic viscosity [m^2/sec]; "u" and "w" – dimensionless velocities.

$Fr = \frac{gh^3}{\nu a}$ Froude number

g is gravitational acceleration

Le = a/D = Sc/Pr - The Lewis number

$Sc = \nu/D$ - The Schmidt number

$\beta = \frac{RT_*}{E}$ - Dimensionless parameter

$\gamma = \frac{c_p RT_*^2}{QE}$ - Dimensionless parameter

$P = \frac{e\sigma K_v (\beta T_*)^3 h}{\lambda}$ - Thermal radiation dimensionless coefficient

$\sigma = 5.67(10^{-8})$ [watt/m^{2}K^4] –Stefan-Boltzmann constant

$K_v = A_o h/V$ – Dimensionless opening factor

$\delta = (\frac{E}{RT_*^2})Qz(\exp(-\frac{E}{RT_*}))$ - Frank-Kamenetskii's parameter

$\bar{W} = \frac{\nu}{h} W$ - Vertical component of composite's velocity

$\bar{U} = \frac{\nu}{h} U$ - Horizontal component of composite's velocity

$b = L/h$, "L" and "h"- Length (width) and height of component accordingly

W; U – dimensionless velocities

Coordinates: $\bar{x} = x / h$ and $\bar{z} = z / h$ - "x" and "z" – dimensionless coordinates.

Chapter 1

Introduction and Assumptions

1.1 Introduction

Without exaggeration, one might say that the cyclic creep-fatigue problem is among the most complex and urgent problems of the mechanics of solid body deformation. It is primarily associated with the development of a creep-fatigue damage mechanism in order to assess the structural stability of composite and nanocomposite structural elements and systems. Today in modern industry, especially where it is critically important to ensure the strength and stiffness of structural elements with minimal weight, the designers of airplanes and airplane engines are using composite and nanocomposite materials [1]. However, existing imperfections in the methods for evaluating the strength of these materials often lead to the fact that large safety factors are required in the structural calculation methods. Therefore, to ensure the required safety of the composite materials' products, the structural engineer designers are using larger margins of safety, which, in turns, reduces their efficiency. These considerations fully apply to methods for evaluating the fatigue strength of composites and nanocomposites. Under high temperature cyclic loading of composite materials, the accumulation of damage can manifest itself in a change in the integral properties of such materials. Various methods of non-destructive testing can detect changes in the elastic modulus, electrical conductivity and damping coefficient. Although the technology of new composite materials is developing rapidly, their use in real structures is being hampered because of the insufficient fatigue data required for design. The stress analysis under variable thermal loads requires taking into account the anisotropy of the so-called averaged properties of the composite. Unlike metals, composite materials are substantially heterogeneous and anisotropic. Damage does not accumulate in a localized form, and failure does not always occur because of the propagation of a single macroscopic crack. Micro structural mechanisms of damage accumulation, including fiber fracture and matrix fracture, fiber-matrix splitting and delaminating sometimes occur independently. At low levels of cyclic loading or in the initial part of the fatigue life, most types of composites accumulate and diffuse the damage. These damages propagate throughout the tension zone, and gradually reduce the strength and stiffness of the composite structure. In the later stages of the fatigue life, the amount of accumulated damage in a certain region of the composite element can be quite large. This leads to the fact that the residual strength of the composite of that region drops below the allowable stress level.

Well before a microstructural understanding of fatigue processes was developed, engineers had developed empirical means of quantifying the fatigue process and designs against it. Perhaps the most important concept is the S-N diagram, where at any given constant cyclic stress amplitude, S, applied to a specimen, the number of loading cycles N_f until the specimen fails is counted. Millions of cycles might be required to cause failure at lower loading levels, so the abscissas of such a diagram are usually plotted logarithmically. Statistical variability is troublesome in fatigue testing, because it is necessary to measure the lifetimes of perhaps twenty specimens at each of the ten or so load levels to construct the S–N curve with statistical confidence. It is generally impossible to apply a cyclic load to a specimen at more than approximately 10 Hz (inertia in components of the testing machine and heating of the specimen often become problematic at higher speeds) and at that speed, it takes 11.6 days to reach 10^7 cycles of loading. Obtaining a full S–N curve is obviously a tedious and expensive procedure. For instance, a very substantial amount of testing is required to obtain a S–N curve for *the simple case of fully reversed loading*, and it will usually be impractical to determine whole families of curves for every combination of mean and alternating stress (see Fig. 1.1) [2].

There are a number of strata gems for assessing the residual stress of fatigue, one common one being the Goodman diagram (see Fig. 1.2) [3]. Here is the graph with mean stress as the abscissa and alternating stress as the ordinate, and a straight “lifeline” is constructed from σ_m on the σ_{alt} axis to the ultimate tensile stress σ_f on the σ_m axis. Then for any given mean stress, the fatigue life limit (the value of alternating stress at which fatigue failure never occurs) can be read directly as the ordinate of the lifeline at that value of σ_m. Alternatively, if the design application dictates a given ratio of σ_m to σ_{alt}, a line is drawn from the origin with a slope equal to that ratio. Its intersection with the “life line” then gives the effective endurance limit for that combination of σ_f and σ_m.

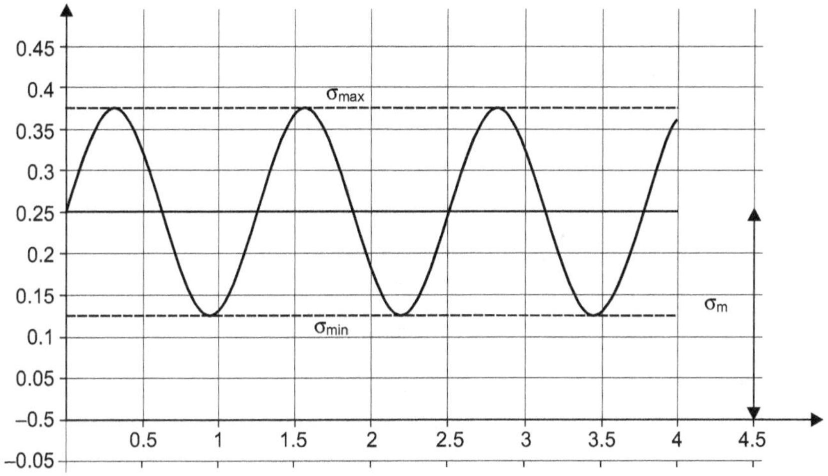

Figure 1.1. Simultaneous mean and cyclic loading.

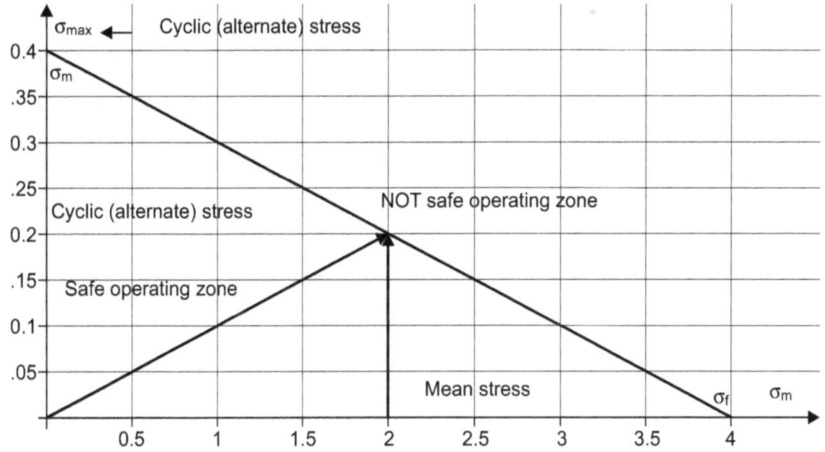

Figure 1.2. The Goodman diagram.

The maximum completely reversing cyclic stress that a material can withstand for an indefinite (or infinite) number of stress reversals is known as the fatigue strength or endurance strength (S_e).

1.2 Fatigue curve and endurance limit

The ability of the material to withstand the effects of variable loads characterizes the endurance limit, the value of which is usually determined experimentally. The purpose of the test is to determine the number of cycles at which each sample fails at a given stress. The first sample is loaded with a

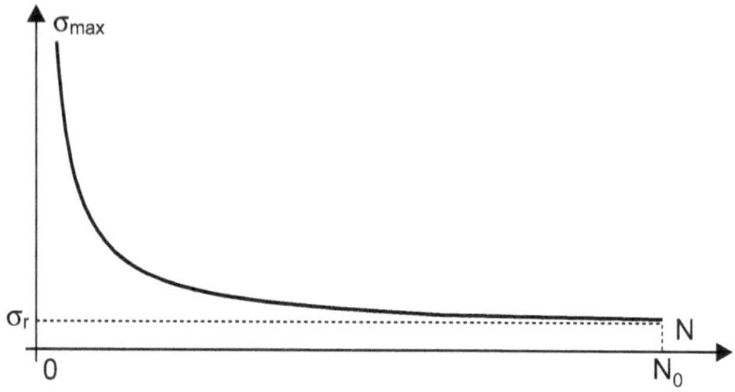

Figure 1.3. Fatigue curve.

symmetric cycle with stress amplitude that brings it to failure. The following samples are tested with reducing loads. The compiled test results are the main source of so-called S–N diagram, which is the fatigue (endurance) curve (see Fig. 1.3) [4]. For many engineering materials, a horizontal asymptote is a characteristic feature of the fatigue curve. This means that at a certain value of the amplitude of the cycle stress, the sample can withstand a theoretically infinitely large number of loading cycles. This stress value is the endurance limit.

For metals, endurance limits are the values at which fatigue failure does not occur after 10^7 cycles. This number of cycles is called basic and is designated N_0. For non-ferrous metals, hardened steels, and many composites, the endurance limit is 20 million cycles, on the fatigue curve. Thermal fatigue is defined as the cracking of a material primarily caused by repeated heating and cooling which induces cyclic internal thermal stresses. In addition, stresses caused by external mechanical loads may also contribute to thermal-fatigue failure. Stress, temperature, and time conditions may be sufficient to cause both instantaneous viscoelastic strain and time-dependent creep during each cycle. Therefore, a general method for calculating the thermal-fatigue life must apply to a cycle in which various types of composite material behaviors are present in arbitrary combinations. It should be noted that the important factor in determining the life span of a composite is the temperature—time dependence of the same type as expressed in creep data, rather than a time-independent effect such as plastic deformation. There are some general assumptions made in this manuscript and they are as follows:

(1) Thermal fatigue is considered as a primary creep-fatigue phenomenon rather than a combination of many factors influencing fatigue failure.

(2) Cyclic creep-fatigue is assumed to have a failure mechanism in any type of loading combination (including "stop – and – go"), rather than being limited to cycles that contain tensile mean stress only.

(3) The proposed analytical method applies directly to the complete service cycles of load and temperature, rather than being limited to a specific idealized laboratory test such as the isothermal strain cycling method.

(4) The well-known cumulative damage continuum models for the creep-fatigue process have some modifications under non-steady loading conditions.

(5) The proposed phenomenological model predicts in quantitative and qualitative terms the effect on the thermal creep-fatigue life of various cycle parameters such as the maximum amplitude of strain (stress); frequency of accelerations; range of temperatures and rates of heating and cooling.

This chapter is devoted to the methods of solving creep-fatigue problems analytically by using the simplest mechanical system (a Standard Linear Model (SLM) or Vought – Kelvin model). Also, the cumulative damage model and the mechanism of application of this model is proposed here. The Vought – Kelvin model was chosen here with the main purpose of confirming the correctness of the proposed assumptions in the formulation of the fatigue problem of composites and nanocomposites under cyclic loading, as well as obtaining a qualitative confirmation of the results. In the subsequent chapters, each feature of cyclic fatigue-creep of composites and nanocomposites under high-temperature loading conditions is considered. Analyses of fatigue-creep of composites and nanocomposites problems are presented with the numerical solutions (in dimensionless form). A uniaxial phenomenological model proposed here takes into consideration an inelastic deformation and failure criteria of composites and nanocomposites under the combined action of static and cyclic loads. The Gibbs energy [5,6] and adjacent thermodynamic [7,8] approaches for describing inelastic rheological deformation and failure of composites and nanocomposites under unsteady loading provides good results, and the expediency of their use in computational practice. The purpose of this work is to generalize the approach [9] that describes the fatigue phenomena that occur in composite materials under the combined action of static σ_{st} and cyclic loads with an amplitude value of the cyclic component σ_a. Consideration is also given to the so-called multi-cycle loading at a frequency $f > 10$ Hz and an amplitude coefficient $a_0 = \sigma_a/\sigma_{st}$ not exceeding a certain nondimensional critical value a_{cr}, having an order of 0.95. In this case, cyclic loading leads to two main effects: (1) acceleration (or even initiation) of the creep process at a given static stress σ_0; (2) a decrease in the accumulated inelastic deformation at the time of failure compared with a similar value under pure static loading. As a rule, these phenomena cannot be described either within the framework of just ordinary classical approaches, or from the standpoint of phenomenological creep only:[7] or from the standpoint of fatigue in a symmetric (or asymmetric) cycle.

1.3 Creep-fatigue process under cyclically changing strain

Dr. R. Talreja [10] gave the most general definition of the creep-fatigue process of composites: "A reliable and cost-effective fatigue life prediction methodology for composite structures requires a physically based modeling of fatigue damage evolution. An undesirable alternative is an empirical approach. A major obstacle to developing mechanistic models for composites is the complexity of the fatigue damage mechanisms, both in their geometry and the details of the evolution process. Overcoming this obstacle requires insightful simplification that allows the use of well-developed mechanical modeling tools without compromising the essential physical nature of the fatigue process". Although the fatigue behavior of composite materials is significantly different from the behavior of metals, many models have been developed based on the well-known S – N curves. These models make up the first class of so-called "fatigue strength models". This approach requires large experimental studies and does not take into account the real mechanisms of damage, such as matrix damage and fiber breaks [11]. The second class includes phenomenological models for multi-cycle fatigue. These models offer an evolutionary law that describes the gradual degradation of the strength or stiffness of a composite sample based on macroscopic properties. Recently, models have been developed based on the concepts of continuum damage mechanics. The development of damage is determined by evolutionary kinetic equations that reflect the irreversible nature of damage [12]. Continuum damage models introduce scalar, vector, or tensor damage parameters that describe the degradation of the entire composite material or structural components. These models are based on physical modeling of the main damage mechanisms that lead to macroscopically noticeable degradation of mechanical properties [13]. The main result of all fatigue models is the prediction of fatigue life, and each of these three categories uses its own criterion to determine the failure process and, as a consequence, the fatigue life of the composite material.

1.4 The concept of effective stresses

In the continuum damage approach the composite material is considered as a homogeneous anisotropic elastic medium. Assuming that the model considers the small strains that take place in the structural element or structural system as a whole, the elastic strain energy function is quadratic and there is a linear relationship between the stress and strain tensors σ and ε. In contrast to the mechanics of brittle fracture, which considers the process of equilibrium or the growth of macrocracks, the continuum mechanics of damage uses continuous internal variables that are associated with the density of microdefects. The proposed model uses the concept of so called effective stress that integrally reflects various types of damage at the micro-scale level (such as the formation and growth of micro cracks of the matrix, fiber breaks, delaminating and other microscopic defects) [14].

1.5 Use of scalar internal variable to quantify damage

The damage parameter ω is associated with a decrease in the effective area of any cross section at a given point on the body and determined by the following formula:

$$\omega = \frac{A - \Delta A}{A} \tag{1.1}$$

In Equation (1.1) "A" is the nominal, undamaged cross-sectional area, "ΔA" is the total cross-sectional area of all defects on this site. By definition, the theoretical value of ω should be in the range $0 \leq \omega \leq 1$. The effective stress tensor in case of isotropic damage is introduced as follows:

$$\tilde{\sigma} = \frac{\sigma_{ij}}{1 - \omega} \tag{1.2}$$

For a more detailed reflection of the mechanism of fatigue damage of composites, tensor measures of damage are used. This allows one to take into account the directed, anisotropic nature of the accumulation of fatigue defects. To identify the parameters of the models of anisotropic damage, a significant number of experiments are required with complex test programs that make it possible to identify the directional nature of fatigue damage.

In considering the geometric interpretation of the anisotropic damage function, the second rank tensor is introduced. For physical reasons, it is symmetrical. Generalizing the dependence Equation (1.2) to the case of the damage tensor of the second rank, the symmetric shape of the effective stresses σ_{ij} is obtained as follows:

$$\tilde{\sigma}_{ij} = A_1 \frac{[\sigma_{ij}]^{n_1}}{[1 - \omega]^{m_1}} + A_2 \frac{[\sigma_{ij}]^{n_2}}{[1 - \omega]^{m_2}} \tag{1.3}$$

1.6 The scalar measure of damage

The introduction of a scalar measure of damage also determines the choice of a mathematical model for describing the accumulation of damage. Under conditions of a high-temperature stress state, the rate of damage accumulation should depend on the joint variants of stress tensors and tensors that determined experimentally characterize the mechanical properties of the composite. The desire to reflect the characteristics of the cycles for each component of the stress tensor leads to a complication of theoretical models. A realistic approach to such situations is to introduce the amplitude of the instantaneous stress a_0 and frequency p. The scalar measure of damage ω is considered as a function that depends on the maximum stress value per cycle; the number of loading cycles N; the cycle parameter R; temperature T; material properties and other variables affecting composite material fatigue. In the proposed model it has been assumed that the damage accumulation rate depends on current stress level and all required material constants are determined from relatively simple experiments at fixed values of the cycle parameter and temperature. The rate of damage level ω is presented as:

$$\frac{d\omega}{d\theta} = f(\sigma, \omega, n_i, m_i) \quad i = \mathbf{1}; \mathbf{2} \tag{1.4}$$

In accordance with the concept of continuum damage mechanics, function f(ω) can be obtained theoretically based on the number of loading cycles or by analyzing changes in the modulus of elasticity with the temperature rise [15]. However, for the practical use of the theory, it is more preferable to identify the functional dependence (1.4) based on the S – N Weller fatigue curves' results [16].

The geometric interpretation of the damage parameter at ω = 1 corresponds to the case when the cross section of the composite material is completely filled with macro-cracks. In practice, the material becomes unstable and collapses when the damage reaches a certain critical value, less than unity. Due to significant nonlinearity of the dependence of the damage parameter on the number of cycles, at the stage preceding the failure, the growth rate increases and tends to infinity. Therefore, the interval of change of the damage parameter is close to unity, in the range $0.9 < \omega < 1$ corresponding to an insignificant change in the ratio of the number of cycles to fatigue life N_f.

The basic experiments to determine the fatigue characteristics of composite materials are experiments on cyclic loading under uniaxial stress conditions. Due to the significant scatter of experimental data, it is necessary to test a large number of samples at various stress levels in order to construct S – N curves. The proposed deterministic and probabilistic models are presented by the examples given below. Moreover, the statistical datasets that are required in order to solve the corresponding probabilistic problems are based on the solutions of the corresponding problems in a deterministic formulation.

The developed model allows predicting fatigue life taking into account the influence of the orientation of the main reinforcement directions relative to the planes of elastic symmetry of the material. The models presented here are using the dimensionless parameters and variables that in turn require the minimum number of material parameters and the minimum necessary set of experiments respectfully.

Until the mid-40s of the last century, the solution to problems of strength of materials and structures under cyclic loading was mainly reduced to the assessment of fatigue resistance formulated first by V. Woehler [17].

1.7 Strength as endurance limit of composites and nanocomposites

The problem of strength under cyclic loading is considered more extensively in present times. This is due to the development of new industries in modern technology, primarily stationary and transport power engineering. It is a very well known fact that in most parts of power machines operating at high temperatures, cyclic loads varying in a wide range of frequencies and amplitudes are imposed on static loads of various kinds. If the structures operate in a nonstationary time mode, then an uneven temperature field with large gradients arises in its elements, causing considerable cyclic temperature stresses. It was found that cyclic loading significantly reduces the creep resistance in the entire frequency range, and the impact of low frequency loads (tenths and hundredths of a hertz) equal to and exceeding the yield strength of the material reduces fatigue resistance. It became obvious that such traditional characteristics of strength as endurance limit, static creep limits and long-term static strength can no longer be considered sufficient criteria for reliable performance. As a result, new directions appeared in the section of high-temperature strength—cyclical creep and long-term cyclic strength. The issues of structural and surface stability of materials, thermal fatigue and stability of structures are also very important. These circumstances led to the creation of new methods and means of determining the resistance of structural materials to deformation and damage development process under cyclic loading, to the creation of appropriate theories and physical models. The efforts of many scientists have already achieved significant progress in the field of theoretical interpretations and the quantitative description of the phenomena of cyclic creep and long-term cyclic strength, and in the field of engineering applications of theoretical results.

Particularly relevant is the intensification of creep by high-frequency cyclic loading applied to the composite and nanocomposite materials, which usually occurs at high temperatures, i.e., in the conditions most characteristic of the operation of many critical parts of modern power machines. These include working blades and disks of gas turbines and compressors, gas pipelines, linings of combustion chambers, fasteners and other parts of structural elements for which the mode of high-temperature multi-cycle loading is one of the main ones. It is also indicative that in composites under high temperatures, cyclic creep-fatigue develops not only at small values of the amplitude of the cyclic component, but also at values exceeding the static load in some cases. The result of the intensification of the creep-fatigue process by a high-frequency load is, as a rule, premature violation of the structural integrity of the part and its failure is a result of an excessively developed residual stresses and failure. In particular, the nature of the fracture of the working blades of gas turbines does not carry purely static or fatigue damage, but indicates damage progression due to cyclic creep-fatigue process.

The problem of cyclical creep-fatigue and durability under multi-cycle loading attracted the attention of researchers around the same period when it became necessary to consider taking cyclic loads into account when evaluating strength. These tasks were set by the needs of the turbine industry, and subsequently arose in the nuclear power industry, chemical engineering, aviation and rocket technology. Most of the known studies in the field of cyclic creep-fatigue are experimental, and the first of them are related to the study of the creep of lead, copper, aluminum, and other pure metals. Recently, however, the more general questions of the cyclic creep-fatigue process that develop under the joint effects of static and cyclic loads have become the focus of attention of researchers [18,19,20]. Due to the high specific stiffness and strength of the fiber-reinforced composite materials, they are used for weight-critical structural applications. However, existing imperfections in the analytical methods for evaluating the strength of composites and nanocomposites often lead to the requirement of large safety factors in the structural calculation methods. Therefore, products of composite materials are often designed with an excessive margin of safety, which reduces the efficiency of their use. These considerations fully apply to methods for evaluating the fatigue strength of composite materials.

1.8 Continuous damage accumulation model

A rational choice of the damage accumulation model allows one to lead to a more efficient use of these composite materials [21]. Under cyclic loading of composite materials, the accumulation of damage can manifest itself in a change in the integral properties that are included in corresponding constitutive equations. By various non-destructive testing methods the change in the elastic modulus, electrical conductivity, and damping coefficient can be obtained. Damage can occur by wave effects and recognized by acoustic emission. Although the technology of composite materials is developing rapidly, their use in real structures is hampered by the lack of access fatigue data, which contains the main operational parameters of structural systems. Unlike metals, composite materials are substantially heterogeneous and anisotropic. Damage does not accumulate in a localized form, and failure does not occur because of the propagation of a single macroscopic crack. Microstructural mechanisms of damage accumulation, including fiber fracture and matrix destruction, fiber-matrix splitting and separation, sometimes occur independently. At low levels of cyclic loading or in the initial condition, most types of composites accumulate and diffuse damage. These damages gradually reduce the strength and stiffness of the composite throughout the stress zone. In the later stages of the fatigue life, the amount of accumulated damage in a certain region of the composite can be quite large. This leads to the fact that the residual composite strength of that region drops to the maximum stress level in cyclic loading and as a result, the failure occurs.

Although the fatigue behavior of fiber-reinforced composite materials is significantly different from the behavior of metals, many models have been developed based on the well-known S – N curves. These models make up the first class of so-called "fatigue strength models". Such an approach requires large experimental studies and does not take into account real damage mechanisms, such as matrix damage and fiber breaks [22]. The second class includes phenomenological models for multi-cycle fatigue. These models offer an evolutionary law that describes the gradual degradation of the strength or stiffness of a composite sample based on macroscopic properties. Recently, new developing models based on the concepts of continuum fracture mechanics are used [23]. In models of this type, damage is quantitatively described by some internal parameters of the material. The development of damage is determined by evolutionary kinetic equations that reflect the irreversible nature of damage. Continuous damage models introduce scalar, vector, or tensor damage parameters that describe the degradation of the entire composite material or for structural components. These models are based on physical modeling of the main damage mechanisms that lead to macroscopically noticeable degradation of mechanical properties [24]. The main result of all fatigue models is the prediction of fatigue life, and each of these three categories uses its own criterion to determine the fracture condition and, as a consequence, the fatigue life of the composite material. In this paper, we consider a continuum model of the accumulation of fatigue damage, based on the assumption that the growth rate of the damage parameter depends on the maximum value of the specific energy of elastic deformation of the composite.

In the continuum approach to the analysis of the stress state and fatigue of products made of composites, the material is considered as a homogeneous anisotropic elastic medium [25]. When constructing the model, it is assumed that small elastic strains take place. The elastic strain energy function is quadratic and there is a linear relationship between the stress and strain tensors σ and ε respectively. In contrast to the mechanics of brittle fracture, which considers the process of an equilibrium state or the growth of macrocracks, the continuum mechanics of damage uses continuous internal variables that are associated with the density of microdefects. The proposed model is based on the concept of effective stress and integrally reflects various types of damage at the micro-scale level, such as the formation and growth of micro-cracks of the matrix, fiber breaks, delamination and other microscopic defects [26]. The complexity and variety of mechanisms for the accumulation of fatigue damage and the degradation of the strength properties of the composite make it justifiable. Damage parameter ω is associated with a decrease in the effective area of any cross section near a given point on the body and is determined by the following formula:

$$\omega = \frac{A - \bar{A}}{A} \tag{1.5}$$

where A is the nominal, undamaged cross-sectional area, $\bar{A}$ is the total cross-sectional area of all defects on this site. By definition, the theoretical value of ω should be in the range $0 \le \omega \le 1$. The effective stress tensor in the case of isotropic damage is introduced as follows:

$$\tilde{\sigma}_{ij} = \frac{\sigma_{ij}}{\mathbf{1} - \omega} \tag{1.6}$$

For a more detailed reflection of the mechanism of fatigue damage of composites, tensor measures of damage are used. This allows one to take into account the directed, anisotropic nature of the accumulation of fatigue damage.

To identify the parameters of models of anisotropic damage, a significant number of experiments are required with complex test programs that make it possible to identify the directional nature of fatigue damage. When considering the geometric interpretation of anisotropic damage, a damage rank tensor of the second rank is introduced. For physical reasons, it is symmetrical. Generalizations of dependence (2) to the case of the damage tensor of the second rank, the symmetric shape of the effective stresses $\tilde{\sigma}_{ij}$ can be obtained as follows [27]:

$$\tilde{\sigma}_{ij} = \frac{1}{2}\left[\frac{\delta_{ki}\sigma_{ik}}{1-\omega_{ki}} + \frac{\delta_{lj}\sigma_{lj}}{1-\omega_{jl}}\right] \tag{1.7}$$

where, δ_{ij} is the Kroniker delta and ω_{ij} is the damage tensor of the second rank.

1.9 Damage accumulation function for composites

The introduction of a scalar measure of damage also determines the choice of a mathematical model to describe the process of damage accumulation. Under complex stress conditions, the rate of damage accumulation should depend on the joint invariants of stress tensors and tensors characterizing the mechanical properties of the composite. The change in the individual components of the stress tensor within the cycle can theoretically occur in different time dependences. The desire to reflect the characteristics of the cycles for each component of the stress tensor leads to an excessive complication of theoretical models. A realistic approach to such situations is to introduce a cycle parameter for the characteristic invariant of the stress state. As such a parameter, the ratio of the minimum and maximum values of the first principal stress per cycle is:

$$R = \frac{\sigma_1^{min}}{\sigma_1^{max}} \tag{1.8}$$

The scalar measure of damage ω considered herein depend on the maximum value of viscoelastic deformation, the number of loading cycles N, the cycle parameter R, temperature T, material properties and other parameters affecting fatigue mode:

$$\omega = f(R, E, \sigma, \theta) \tag{1.9}$$

Establishing the dependence of fatigue strength on the cycle parameter and temperature is a difficult problem even for homogeneous materials. Material constants of composites and nanocomposites are usually determined from relatively simple experiments at fixed values of the cycle parameters and temperature. In the proposed model, the hypothesis that the rate of damage accumulation depends on the current stress value, the ratio of the minimum stress to the maximum R and the current damage level ω is accepted.

$$\frac{d\omega}{dt} = f(R,\sigma,\theta) \tag{1.10}$$

The form of the function f, which determines the rate of damage accumulation is usually obtained based on the results of experiments on fatigue strength. In accordance with the concept of continuum damage mechanics, function g(ω) can theoretically be controlled with temperature—time (the number of loading cycles) by a change in the elastic modulus. For practical use of the theory, it is preferable to identify the functional dependence (6) based on the results of fatigue strength from S – N Weller curves [28].

1.10 Paris' Law and Miner's Rule

The effective stress range using a rain flow cycle counting algorithm and Miner's linear cumulative damage law is as follows [29,30]:

$$\sum \frac{n_i}{N_i} = 1 \tag{1.11}$$

where n_i is the number of cycles at stress range σ_i and N_i is the constant amplitude fatigue life at the stress level σ_i. These values can be rewritten as:

$$N_i = A\sigma_{r_i}^{-n} \tag{1.12}$$

and $n_i = \gamma_i T$, where T is the total number of cycles and $\gamma\, \gamma_i$ is the fraction of the total cycles occurring at stress level σ_i. Substituting these values into Miner's rule yields:

$$\sum \frac{n_i}{N_i} = \sum \frac{\gamma_i T}{A\sigma_{r_i}^{-n}} = \frac{T}{A}\sum \gamma_i \sigma_{r_i}^{n} \quad \text{or:}\ T = \frac{A}{\sum \gamma_i \sigma_{r_i}^{n}} \tag{1.13}$$

Equation 5 has the same form as Equation 4, the equation for the constant amplitude fatigue life. The effective stress range, defined as the constant amplitude stress range that will produce the same fatigue life as the variable amplitude stress history, is:

$$\sigma_{re} = \left(\sum \gamma_i \sigma_{r_i}^{n}\right)^{1/n} \tag{1.14}$$

This relationship was used to relate the typical traffic vehicle spectrum of fatigue stresses into a constant level stress range that produces the same fatigue damage.

As introduced in the abovementioned, fatigue in structures is the process of growth of cracks under the action of repetitive tensile loads. The fatigue life of a component is dependent upon the applied stress range, the initial discontinuity introduced during fabrication, and local stress increase due to the joint geometry. Below is an example for the application of Miner's rule to a variable spectrum of amplitudes (see Table 1.1 below).

This example applies the concepts of Miner's rule to determine the effective dimensionless stress range, 1.645, and provides an assessment of which cycles contribute most and least to the total fatigue damage. Of note is that the stress ranges from 0.5 to 1.5 constitute 80 percent of the total number of cycles but cause only 16 percent of the total damage. The highest three stress ranges, 3.5 to 4.5, only constitute 3 percent of the total number of cycles but cause a much larger amount of damage (41 percent)! This is consistent with life, or damage related to the cube of the stress amplitude ranges, with a small percentage of high stress ranges using up a large portion of the structural fatigue life details available .

In 1945, M. A. Miner popularized a rule that had first been proposed by A. Palmgren in 1924. The rule is variously called Miner's rule or the Palmgren-Miner linear damage hypothesis

If $\theta_{max} = 10.25$ then $N_f = (10.25)(10^{4.05})/(2\pi) = 183045$ cycles

$$E_0 = 1[\text{GPa}];\ \alpha = 10^{-4}[1^0\text{K}];\ \sigma_f = 10.25(20)(10^3)(10^{-4}) = 20.5[\text{MPa}] = 2.97[\text{ksi}]$$

$$T_* = 600^0\text{K};\ T_{max} = \beta T_*\theta + T_* = 0.0333(600)10.25 + 600 = 20(10.25) + 600 = 805^0\text{K}$$

$$\sigma_{max} = 0.175[\text{GPa}] = 175[\text{MPa}] \gg 20.5[\text{MPa}]$$

The formulation, based on thermodynamic principles, leads to a system of kinetic equations for the evolution of damage. An effective viscosity inversely proportional to the rate of damage increase is introduced to account for

Table 1.1. Application of Miner's rule to a variable spectrum of amplitudes.

Stress interval	**Percentage of amplitudes γ_i (%)**	**Amplitude σ_{0i}**	**Amplitude $(\sigma_{0i})^3$**	**Relative damage $\gamma_i(\sigma_{0i})^3$**	**Percent Damage**
1	40	0.5	0.125	0.05	1
2	25	1.0	1.0	0.25	5
3	15	1.5	3.375	0.50625	10
4	10	2.0	8.0	0.8	16
5	5	2.5	15.625	0.78125	16
6	2	3.0	27	0.54	11
7	1	3.5	42.875	0.42875	9
8	1	4.0	64.0	0.64	13
9	1	4.5	91.125	0.91125	19
Total	100%			Total: 4.9075	100%
Effective stress				$\sigma_{eff.} = (4.45185)^{1/3} = 1.645$	

Table 1.2. Palmgren-Miner linear damage.

frequencies @ stress level	Percentage of frequencies γ_i (%) per fatigue life span L = 100%	log p_i	$(\log p_i)^3$	Relative damage $\gamma_i(\log p_i)^3$	Percent Damage
$p = 1$ @ $\sigma_{0i} = 0.5$	1	0	0.0	0.0	0
$p = 10$ @ $\sigma_{0i} = 1.0$	15	1.0	1.0	0.15	5
$p = 10^2$ @ $\sigma_{0i} = 1.5$	25	2.	8	2.0	10
$p = 10^3$ @ $\sigma_{0i} = 2.0$	20	3.0	27	5.4	16
$p = 10^4$ @ $\sigma_{0i} = 2.5$	15	4.0	64	9.6	15
$p = 10^5$ @ $\sigma_{0i} = 3.0$	10	5.0	125	12.5	11
$p = 10^6$ @ $\sigma_{0i} = 3.5$	10	6.0	216	21.6	8
$p = 10^7$ @ $\sigma_{0i} = 4.0$	3	7.0	343	10.29	13
$p = 10^8$ @ $\sigma_{0i} = 4.5$	1	8.0	512	5.12	18
Total	100%			Total: 66.66	100%
Effective frequency				$\log p_{eff.} = (66.66)^{1/3} = 4.05$ $p = 10^{4.05}$	

gradual accumulation of irreversible deformation due to dissipative processes. A power-law relation between the damage variable and elastic modulus leads to a non-linear coupling between the rate of damage evolution and the damage variable itself.

1.11 The Bergman-Milton theory

If the loading is a sequence of constant stresses with hold times and unloading or if it is a cyclic loading, the phenomenon of fatigue occurs as a superimposed effect. It is called creep-fatigue interaction because the earliest models were a combination of two terms: one for creep, and one for fatigue with or without couplings to obtain a model of nonlinear interaction.

The simple Taira rule of linear interaction (1962) applies to isothermal uniaxial cyclic loading with $0 \leq \sigma \leq \sigma_{max}$ and with a hold time of Δt. If N_R is the number of cycles to rupture corresponding to a time $t_R \approx \Delta t \cdot N_R$; if N_{RF} is the number of cycles to rupture in pure fatigue ($\Delta t \approx 0$) for the same stress range; and if t_{Rc} is the time-to-rupture in pure creep for the same maximum stress $\sigma = \sigma_{max}$ for all t, then the Taira rule reads [31,32]:

$$\frac{t_R}{t_{RF}} + \frac{N_R}{N_{RF}} = 1 \tag{1.15}$$

The time-to-rupture t_R at constant stress and constant temperature is the solution of this differential equation for $\dot{\omega} = f(\sigma, \omega)$ with the initial condition $t=0 \rightarrow \omega=0$

$$\dot{\omega} = \left[\frac{\sigma}{A_{\omega}(1-\omega)} \right]^r \tag{1.16}$$

where, A_{ω} and "r" are material parameters

A composite is an inhomogeneous structure. However, it can be characterized as homogeneous by using effective parameters, obtained through averaging, or homogenization, if the sizes of inhomogeneities are much smaller than the characteristic length of a bulk material. Any existing homogenization theory tries to ascribe effective parameters to a mixture of different phases, providing mixing rules. A mixing rule is an analytical formulation that describes an effective parameter as a function of the size of inclusions and concentration. There is currently a multitude of mixing rules in the literature based on different homogenization theories and approaches. They are applicable to different types of mixtures, rates of generalization, and depend on which parameters are homogenized, and what their limits are.

1.12 Failure criteria

In most studies, two-component mixtures are considered, where identical inclusions are embedded in a homogeneous matrix. Effective properties of such a composite depend on the intrinsic properties of the inclusions and the matrix, as well as on the morphology of the composite. The morphology is a characterization of the manner, in which inclusions are distributed in the composite, including their concentration, shape, and correlations in the location. Therefore, the morphology determines how inclusions are shaped and distributed, whether they are mutually aligned/misaligned in the composite, and what concentrations of inclusion phases and a matrix material are. A conventional approach to describe the properties of composites employs mixing rules, i.e., equations that relate the intrinsic properties of inclusions and the matrix with the effective properties of the composite based on a simple idealized model considering an ellipsoidal-shaped inclusion. Typically, the characterization of the concentration and the shape of inclusions are included explicitly in the mixing rules, and the account for other morphological characteristics is attempted by a proper selection of the mathematical form of mixing rules. A number of mixing rules are found in the literature. The basic mixing rules are the Maxwell Garnet equation (MG) [33], Bruggeman's Effective Medium Theory (EMT) [34], and the Landau-Lifshitz-Looyenga mixing rule (LLL) [35].

Effective medium approximations (EMA) pertain to analytical modeling that describes the macroscopic properties of composite materials.

The Maxwell Garnett equation is as follows:

$$\frac{\psi_{eff} - \psi_m}{\psi_{eff} + 2\psi_m} = \varphi\left(\frac{\psi_{incl} - \psi_m}{\psi_{incl} + 2\psi_m}\right) \tag{1.17}$$

where ρ_{eff} is the effective constant of the medium, ρ_{inc} of the inclusions, and ρ_m of the matrix; φ is the volume fraction of the inclusions ρ_{inc}.

Bridgeman's model

Without any loss of generality, we shall consider the study of the effective for a system made up of spherical inclusions with different arbitrary values. Then the Bridgeman formula takes the form:

$$\frac{\psi_m - \psi_{eff}}{2\psi_{eff} + \psi_m}(1-\varphi) + \varphi\left(\frac{\psi_{incl} - \psi_{eff}}{\psi_{incl} + 2\psi_{eff}}\right) = 0 \tag{1.18}$$

The Landau–Lifshitz–Looyenga mixing rule (LLL) is written as

$$(\psi_{eff}+1)^{1/3} - 1 = \varphi\left((\psi_{incl}+1)^{1/3} - 1\right)$$

$$\frac{\psi_{eff}}{1+n\psi_{eff}} = \varphi\frac{\psi_{incl}}{1+n\psi_{incl}}$$

$$\frac{1}{n\psi_{eff}} = \frac{1+n(1-\varphi\psi_{incl})}{n\varphi}\psi_{incl} \tag{1.19}$$

$$\psi_{eff} = \frac{\varphi\psi_{incl}}{[1+n(1-\varphi)\psi_{incl}]}$$

The LLL expression was extensively used to describe the properties of dispersive systems composed of powders or exhibiting porous structures. It was even shown that the LLL formula was more reliable when mixtures contained strongly dissipative particles and was compared to others like Maxwell Garnett (MG) and Bruggeman,

$$d_{eff} = (1+\varphi((\theta+1)\text{^}0.333-1))\text{^}3 - 1 \tag{1.20}$$

Equation (1.20) is written for the generalized viscosities of inclusions, ψ_{incl}, and the effective viscosity, ψ_{eff}, both normalized to the viscosity of the matrix. The LLL mixing rule is built up by an iterative procedure starting from a homogeneous material of inclusions and replacing a small amount of this material by the material of the matrix. After that, the resulting "effective" material is regarded as the homogeneous component for the succeeding substitution step, and so on, which results in Equation (3). The mixing rule obtained by the same iterative procedure starting with the homogeneous matrix is referred to as the asymmetric Bruggeman approximation. The result of the LLL mixing rule is independent of the form factor of inclusions. The LLL mixing rule is known to be an accurate result for the case when the material parameter of inclusions differs slightly from that of the matrix. Agreement of both the MG and EMT mixing rules with the LLL mixing rule in the case of the susceptibility of inclusions slightly differing from zero is attained only if $n = 1/3$. When the volume fraction of inclusions is small, $p << p_c$, and the interaction between the inclusions is negligible, all three theories are reduced to:

$$\psi_{eff} = \frac{\varphi \psi_{icl}}{1 + n\psi_{icl}} \tag{1.21}$$

Strictly speaking, Equations (1.18), (1.19), and (1.20) are valid for the case of perfectly spherical shaped inclusions, which have a shape factor n equal to 1/3. For non-spherical particles, the inclusions must be averaged over all three principal axes of the inclusion (1.21). Two particular cases of non-spherical inclusions are of practical interest—nearly spherical inclusions and highly elongated inclusions (long fibers or platelets). For nearly spherical inclusions, the composites are conventionally described by Equations (1.18), (1.19) and (1.20) involving an averaged form factor , which is found empirically and may differ from 1/3. For elongated inclusions, the form factor along the shorter axis (in the platelet case), or the sum of two form factors along the shorter axes (in the fiber case) is close to unity, and the orientation of inclusion in other directions can be neglected. In this case, the above equations are valid again, with a randomization factor, κ, included in the right-hand part of the equations to account for an alignment of non-spherical inclusions. For a fiber-filled composite, $\kappa = 1/3$, when the fibers are randomly oriented in space, and $\kappa = 1/2$, when the fibers are randomly oriented in a plane. For composites filled with platelet-shaped inclusions, $\kappa = 2/3$ for the 3D isotropic orientation.

1.13 The Bergman-Milton theory

A generalization of mixing rules may be made with the use of the Bergman-Milton spectral theory (BM). The theory expresses the effective material parameter of a composite as

$$\chi_{eff} = fi\int_0^1 \frac{b(n)}{1+n\chi_{icl}}dn$$
$$\int_0^1 b(n)dn = 1 \text{ and } \int_0^1 nb(n)dn = \frac{1-fi}{D} \qquad (1.22)$$

where the spectral function, b(n), is introduced as a quantitative characterization of the composite's morphology. One can see from Equation (1.22), the BM theory accounts for a distribution of effective form factors of inclusions in a composite. This distribution may be associated with the following statistical parameters and processes within the composite: a spread in the shapes of individual inclusions comprising the composite; possible agglomeration of inclusions to clusters; and the effects of multiple scattering and inhomogeneous fields affected by neighboring inclusions. Again, the spectral function is the same for all susceptibilities of a particular composite. The sum rules relate the spectral function b(n) to the volume fraction of inclusions fi for a macroscopically isotropic composite in D dimensions. The practically important cases are D=3 (an isotropic 3D composite with non-aligned randomly distributed inclusions, the shape of which is arbitrary in the general case) and D=2 (an assembly of infinitely long cylinders). The sum rules provide an agreement of the spectral theory with the LLL mixing rule at $\chi_{incl}\rightarrow 0$.

A well known example of such a combination is the Lichtenecker mixing rule [36], which is written as:

$$\psi_{eff}^{k} = \varphi\psi_{incl}^{k} + (1-\varphi)\psi_{matrix}^{k} \qquad (1.23)$$

In Equation (1.23), k has a physical meaning of a critical exponent, which is conventionally treated as a fitting parameter to obtain an agreement with measurements. Equation (1.23) may be considered as an empirical combination of the LLL mixing rule and the introduction to long range correlations.

This kind of test may be helpful in the validation of the unified damage law for creep-fatigue damage if the strain to rupture ε_{pR} and the strain at the inflexion point corresponding to the threshold $\varepsilon_{p\omega}$ are recorded. The integration of the differential law for σ=const is particularly easy.

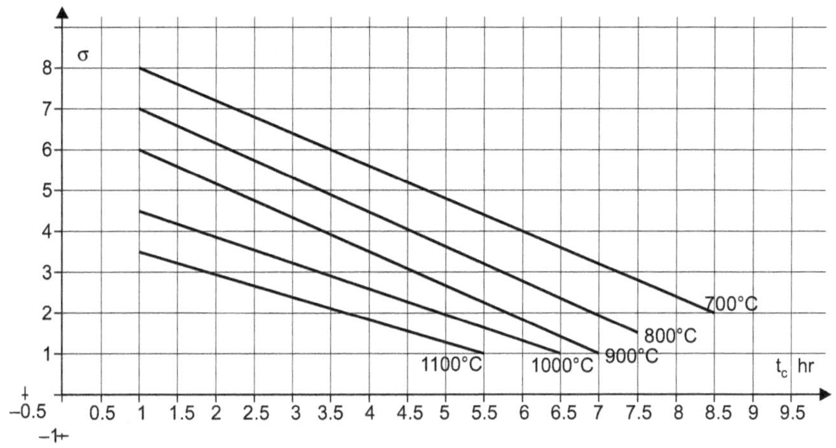

Figure 1.4. Isochronous curves of super alloy IN 100 at different temperatures.

The most important tasks here are related to determining the conditions for the intensive development of cyclic creep, evaluating the interaction of fatigue and creep, as well as obtaining quantitative relationships between deformation, cyclic creep rate, temperature and stress of the cycle. Problems of analytic estimation of the limiting state of materials under conditions of developed cyclic creep are of considerable interest. To describe the experimentally obtained high-temperature curves of creep and fracture defining equations like Bailey-Norton (the time-hardening formulation of power law creep) [37] and Rabotnov-Kachanov [38] are used:

$$\dot{\varepsilon} = B\frac{\sigma^n}{(1-\omega)^m};\quad \dot{\omega} = D\frac{\sigma^n}{(1-\omega)^m} \tag{1.24}$$

$$\omega(0) = 0; \omega(t^*) = \omega_2$$

It is assumed that for a certain period of time Δt after the application the load matrix as a result of creep gains has a deformation increment $\Delta\varepsilon_m = f(\sigma_m, T)\Delta t$. From the condition of compatibility deformation the same increment gets the filler $\Delta\varepsilon_f = \Delta\varepsilon_m$. An increase in fillers' strains leads to an increment in of tensile stresses $\Delta\sigma_f = E_f \Delta\varepsilon_f$ in them and, in turn, reduction of stresses in the matrix according to the additivity rule.

$$\sigma_m = \frac{\sigma_c - \sigma_f V_f}{1 - V_f} \tag{1.25}$$

Since the total load on the composite is $P_c = \sigma_c F_c = \text{const}$, where F_c is the cross section of a composite (assumed unchanged in the process of creep of the composition). The use of the additivity equation is based on the assumption of uniform distribution of deformations in the matrix. By virtue of small volume fractions of fibers ($V_f \approx 5\%$) in the studied material this assumption is justified. However, in general, the principal of taking into account the uneven distribution of deformations and stresses in components is also important.

In the next time step Δt, the creep of the matrix will occur at lower stresses σ_t and thus is modeled as damped creep of the matrix material characterized by a decrease in stresses in it and elastic reloading from the reinforcement.

Reinforcement of composites involves scattering of strength properties and this results in the progression of final failure of the material . The accumulation of cracks in the composite leads to a decrease in their carrying capacity. Assuming that the accumulation of discontinuities occurs statistically uniformly in the region of fragmentation of fibers, we introduce the damage accumulation function W (σ_f), which characterizes the ratio of the number of failed fibers in a layer of critical length equal to the fragment length to the total number of fibers. As the ruptures accumulate, the carrying capacity of the composite decreases and the stresses are redistributed from the destroyed fibers to the matrix.

Exponential Model of viscosity

An exponential model for the temperature-dependence of shear viscosity (μ) was first proposed by Reynolds in 1886 [39]: $\mu(T) = \mu_0 \exp(-bT)$, where T is the temperature and μ_0 and b are coefficients. This is an empirical model that usually works for a limited range of temperatures.

Arrhenius model

This model is based on the assumption that the fluid flow obeys the Arrhenius equation for molecular kinetics [40]: $\mu(T) = \mu_0 \exp\left(\frac{E}{RT}\right)$, where T is the temperature, μ_0 is a coefficient, E is the activation energy and R is the universal gas constant. A first-order fluid is another name for a power-law fluid with an exponential dependence of viscosity on the temperature.

Williams-Landel-Ferry model

The Williams-Landel-Ferry model, or WLF for short, is usually used for polymer melts or other fluids that have a glass transition temperature. The model is: $\mu(T) = \mu_0 \exp\left(\frac{C_1(T - T_g)}{C_2 + T - T_g}\right)$, where T-temperature, C_1, C_2, T_g and μ_0 are empirical parameters (only three of them are independent from each other). If one selects T_g—the glass transition temperature, then the parameters C_1, C_2 become very similar for the wide class of polymers. Typically, we get $C_1 = 17.44$ K and $C_2 = 51.6$ K.

1.14 Failure criteria

Progressive damage models often employ a failure criterion along with degradation models. Different failure criteria have been proposed in the literature such as the maximum stress criterion, Haskin's criterion [41] and Puck's criterion [42].

Maximum stress theory

The maximum stress theory predicts failure when the stresses in the principal material axes exceed the corresponding material strength. In order to avoid failure, it has to be ensured that the stress limits are not exceeded,

$$
\begin{aligned}
&-R_{\Updownarrow}^{c} < \sigma_{\Updownarrow} < R_{\Updownarrow}^{t} \\
&-R_{\Leftrightarrow}^{c} < \sigma_{\Leftrightarrow} < R_{\Leftrightarrow}^{t} \qquad (1.26) \\
&|\tau|_{\lrcorner} < R_{\lrcorner}
\end{aligned}
$$

As soon as one of the inequalities above is violated the material fails by a failure mode that is associated with the allowable stress. This failure criterion does not take any interaction of stress components into account and therefore certain loading conditions such as superposition of tensile and shear stresses, lead to non-conservative results.

Haskin's failure theory

One of the first failure criteria applied to fatigue distinguishing fiber-failure and matrix-failure modes was proposed by Hashin and Rotem. They derived the fatigue failure criterion from their formulation under static loading, which is stated in fiber mode as:

$$
\begin{aligned}
&\sigma_1 = R_{\Updownarrow}^{t} \ \text{ for } \sigma_1 \geq 0 \\
&|\sigma_1| = R_{\Updownarrow}^{c} \ \text{ for } \sigma_1 < 0 \\
&\text{and for the 2D inter-fibre failure:} \qquad (1.27) \\
&\left(\frac{\sigma_2}{R_{\Updownarrow}^{t}}\right)^2 + \left(\frac{\tau_{12}}{R_{\lrcorner}}\right)^2 = 1 \ \text{ for } \sigma_2 \geq 0 \\
&\left(\frac{\sigma_2}{R_{\Updownarrow}^{c}}\right)^2 + \left(\frac{\tau_{12}}{R_{\lrcorner}}\right)^2 = 1 \ \text{ for } \sigma_2 < 0
\end{aligned}
$$

In case of inter-fiber failure Hashin and Rotem proposed an elliptic equation, which depends on the transverse stress σ_2 and on the in-plane stress τ_{12}.

Phenomenological models include a description of the damage in composites during fatigue loading where gradual deterioration of macro- and meso-mechanical properties is described through residual stiffness or strength models. A phenomenological model based on residual strength uses experimental observations to describe the strength loss of composites and is subdivided into two models: sudden death model and wear-out model [43]. The residual strength in the sudden death model is kept constant over a certain number of cycles and is then suddenly degraded drastically when it reaches the critical number of cycles to failure N_f.

This model is a suitable technique to describe this behavior for high-strength unidirectional composites subjected to a high level state of stress [44]. The residual strength in the wear-out model is contrarily continually decreasing over the number of cycles following a certain predetermined equation. These models can be utilized at lower level states of stress. In the wear-out model according to Halpin et al [45], it is assumed that the residual strength R(n) is a monotonically decreasing function of the number of cycles n, and that the change of the residual strength can be approximated by a power-law growth equation:

$$\frac{R^m(n)-\sigma^m}{R^m(0)-\sigma^m}=1-\frac{n}{N_f} \tag{1.28}$$

as reformulated by Shokrieh and Lessard [46] with m as a constant depending on the material. This model has been used by many authors in probabilistic and mechanistic models.

The basis of the concepts of "catastrophic" failure considered below is based on the physical model of damage accumulation in the process of creep deformation. In no way rejecting directions in the study of micro-stresses and micro-deformations, it should nevertheless be noted that this kind of theory at the level of micro-inhomogeneous media are quite complicated even in the case of a uniaxial stress state and therefore are not well suited for solving, for example, boundary problems of continuum mechanics. For these purposes, more preferred are the usual phenomenological theories of creep at the macro level.

Previous phenomenological creep-fatigue models devoted to life prediction and damage assessment were mainly developed on the basis of the Manson-Coffin law [47]. The effects of all these mechanisms are summarized under the label creep damage in phenomenological theory. The term "damages" designate the general destructive effect of the various changes occurring in the material, which eventually limit its usefulness. Thus, the visible damage may consist of an accumulation of permanent deformation from creep (viscous flow), plastic flow, initiation of corrosion pits, or development of macrocracks. These visible effects are the result of submicroscopic changes in the crystalline lattice or in the intercrystalline boundary material. The processes designated by the terms slip, diffusion, recrystallization, rotation of grains, phase change, and relaxation involve varying degrees of atomic rearrangement that take place within the material; any of these processes may constitute the initial stages of "damage".

A very successful suggestion for the description of damage is due to Kachanov [48]. His leading idea is that only a part of the cross-section 'A' of the creep sample has to bear the force, namely the effective or reduced cross-section A_e, reduced by cracks, pores and more. This leads to effective stresses

$$\sigma_{ef} = \frac{F}{A_e} > \frac{F}{A} \tag{1.29}$$

If the effective stresses enter the creep law, this causes a growth in the creep rate. For this aim one introduces a damage parameter ω as the ratio,

$$\omega = \frac{A - A_e}{A} \tag{1.30}$$
$$0 < \omega < 1$$

1.15 Principle of stress equivalence

The damaged material behaves like the undamaged one if the stress in the creep law is substituted by the effective stress [49].

The damage parameter ω is considered as an internal variable, for which one needs an evolution equation. By analogy to the Burgers Standard model, one may write the equation as follows

$$\dot{\omega} = -\alpha\dot{\sigma}_{ef} + \beta\sigma_{ef} + \gamma\omega \tag{1.31}$$

The rate of the effective stress is then,

$$\dot{\sigma}_{ef} = \frac{F}{A_e}\left(\frac{F}{A_e}\right)' = \frac{F'}{A_e} - \frac{F}{A_e^2}A_e' \tag{1.32}$$

In a monotonous creep test, the applied load F is constant, so that F'=0. The growth of the effective cross-section area is fed by different sources after substituting Equation 1.32 and Equation 1.33 into Equation 1.31.

$$\begin{aligned} \dot{\sigma}_{ef} &= -\frac{F}{A_e^2}A_e' \\ \dot{\omega} &= \alpha\frac{F}{A_e^2}A_e' + \beta\frac{F}{A_e} + \gamma\omega \end{aligned} \tag{1.33}$$

In this manuscript, we introduced the function ω, which is the area of cracks per unit cross-sectional area at a given time. In the case of this model, the crack rate of area growth is equal to the crack growth rate v, that is,

$$\partial\omega/\partial t = \omega^{\beta}\, v(\sigma) \tag{1.34}$$

If $\beta = 1$, the first derivative $d\omega/dt = \omega v(\sigma)$. The average stress in the cross section at normal stress is $\sigma = \sigma_0/(1 - \omega)$. The latter formula expresses the effect of increasing the stress level in the damaged material, since $0 < \omega < 1$. Introducing the expression $\sigma = \sigma_0/(1 - \omega)$ into equation (1.1), we can obtain a differential equation for function ω (t), where σ_0 is a given function of time. At the initial time $t = 0$, the function ω is zero, $\omega = 0$, at the failure time $t = T \rightarrow \omega = 1$.

The equation can be easily integrated in the case when the cumulative damage growth rate is a power function of the stress $v = A\sigma^{\alpha}$. In this case, the variables are separated and are as follows $(1 - \omega)^{\alpha}\, \omega^{-\beta} d\, \omega = A\sigma dt$. The time fraction rule can be used to predict creep-fatigue life with the help of the following relation,

$$\sum_{i=1}^{m}\left(\frac{n}{N_f}\right)_i + \sum_{j=1}^{m}\left(\frac{\sum t_h}{T_r}\right)_j \le 1 \tag{1.35}$$

where, n is the number of cycles to failure in creep-fatigue, N_f is the number of cycles to failure in pure fatigue for the given strain range, th is the hold time in each cycle, Tr is the time to rupture in pure creep for a given stress and temperature; and ω is the allowable combined damage fraction. In this study N_f was found from a fatigue strain-life curve and T_r was calculated from a Larson-Miller plot. The first term in Equation (1.35) is a cycle fraction representing fatigue damage and the second term is a time fraction representing creep damage.

Phenomenological uniaxial energy type models are proposed to describe the inelastic creep deformation and fracture of composites as the joint action of the static and cyclic loads. The energy and adjacent thermodynamic approaches for describing inelastic rheological deformation and failures modes of composites in conditions of unsteady thermal loading are very important, and the expediency of their application in the structural calculation practice is beyond doubt. The aim of this book is to generalize the author's approach [50] to describe the class of phenomena occurring in composite and nanocomposite material under combined action of static σ_0 and cyclic loads with the amplitude of the cyclic component being σ_a. It is confined here to the consideration of the so-called low-cycle loading at a frequency f > 10Hz and the amplitude coefficient $A = \sigma_0/\sigma_a$, which does not exceed a certain critical value A_{cr}. In the case under consideration, cyclic loading leads to two main effects [51]: (1) acceleration (or even initiation) of the creep process for a given static stress σ_0; (2) reduction of the accumulated inelastic deformation at the moment of failure in comparison with the analogous value under purely static loading. This process is called cyclic creep or cyclic—creep-fatigue. These phenomena cannot be described either within the framework of conventional classical approaches, nor from the point of view of phenomenological creep, nor from fatigue positions in the asymmetric cycle [52]. From the analysis of works on cyclic creep-fatigue it follows that at the phenomenological level, the following approaches can be conventionally distinguished.

1. The introduction of a reduced stress equal to a static stress at which the durability in the static creep mode coincides with the durability in the cyclic creep mode. In this approach, the similarity of the curves of static and cyclic creep is postulated, which is one of its drawbacks. In addition, under non-stationary loading regimes such theories give large errors in both the static and cyclic components. Numerous attempts to create a universal principle of nonlinear summation of damage to date have not led to success.
2. The basic model for the solution of the problem posed in the present manuscript is similar to the model proposed for a quasistatic loading.
3. Below is a procedure for determining the parameters of the proposed model.

   With the energy approach, the following experimental datasets are used as the base:

   - diagram of material stretching at a constant sufficiently high rate of deformation;
   - series of creep curves from the beginning of loading to the moment of failure at σ_0 = const, which are called stationary creep curves; series of creep curves from the beginning of loading to the moment of failure at σ_0 = const, σ_a = const, called the stationary cyclic creep curves.

With the introduction of effective stress, isotropic damage and using the principle of deformation equivalence [53–55], the uniaxial differential type constitutive creep-fatigue equation has a form.

1.16 Standard linear model

Differential equation in this case is as follows:

$$E n\dot{\varepsilon} + H\varepsilon = n\dot{\sigma} + \sigma$$

$$E = E_1; H = \frac{E_1 E_2}{E_1 + E_2}; n = \frac{\eta}{E_1 + E_2} \tag{1.36}$$

If, for example, $E = E_1 = E_2$ and $\varepsilon = \varepsilon_0\,[\sin(pt)]$, then the dimensionless form of Equation 1.36 is as follows:

$$E = E_1; H = \frac{E}{2}; n = \frac{\eta}{2E}; \varepsilon = \varepsilon_0 \sin pt$$

$$\frac{d\sigma}{d\theta} = -\frac{1}{n} m l \sigma + \frac{E}{2n} m l + E$$

$$E = E_0\,[(0.625 - 0.375*\tanh(c*(\theta - \theta_g)))]$$

$$\theta = \frac{E_a}{RT_*^2}[T - T_*];\ \beta = \frac{RT_*}{E_a};\ \tau = \frac{a}{L^2}t \tag{1.37}$$

a -thermal diffusivity of composite materials

L - linear dimension ; T_* - Base Temperature [^{0}K]

θ_g – glass trunsional temperature (dimensionless)

For purposes of comparison with all subsequent cases of fatigue creep in composites we shall first consider the example of pure creep under high temperature influence.

Example 1.1a—pure creep failure (w/o harmonic elongations and Damage function $\boldsymbol{\omega}$)

If $E \neq$ const. [$E = (0.625 - 0.375*\tanh(c*(\theta - \theta_g)))$]; $\sigma_{max}/\sigma_0 \approx 75$ and n = const. then Equation (1.2) has numerical solution (using POLYMATH software [56]):

Data: $E_0 = 0.583(10^4)$ MPa; $\varepsilon_0 = 0.3(10^{-4})$; $c = 5$; $\theta_g = 4$

Calculated values of DEQ variables

	Variable	Initial value	Minimal value	Maximal value	Final value
1	E	1.	0.25	1.	0.25
2	m	0	–0.1895272	0.0705855	–0.1895272
3	m1	0.0405	–0.1400365	0.0405	–0.1400365
4	n	0.1	0.1	0.1	0.1
5	t	0	0	13.5	13.5
6	Y	0.175	0.175	12.70131	12.70131

Differential equations

1 d(Y)/d(t) = –(1/n)*Y*m1 + 0.175*E +0.175* 0.5*E*(1/n)*m1

Explicit equations

1 m = (0.0405*t – 0.01126*t^2 + 0.001462*t^3 – 0.00006868*t^4)

2 m1 = (0.0405 – 0.02252*t^1 + 0.004386*t^2 – 0.0002747*t^3)

3 E = (0.625 – 0.375*tanh(5* (t – 4)))

4 n = 0.1

Model: Y = a1*t + a2*t^2 + a3*t^3 + a4*t^4 + a5*t^5

Variable	Value
a1	1.110151
a2	–0.7023468
a3	0.1754299
a4	–0.0181235
a5	0.0006566

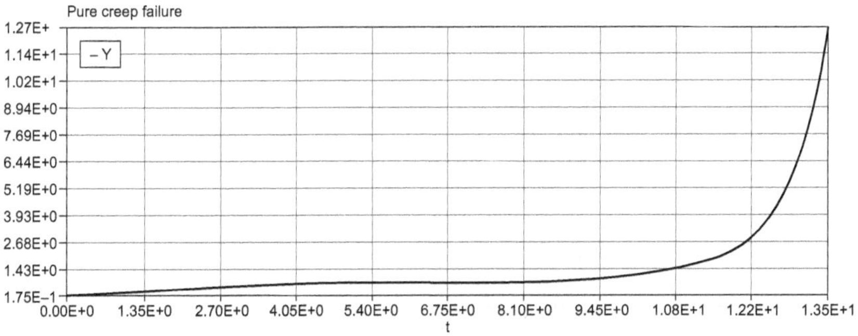

Figure 1.5. Pure creep failure.

$$\sigma = 1.11\theta - 0.70\theta^2 + 0.175\theta^3 - 0.018\theta^4 + 0.000657\theta^5 \quad (1.38)$$

$\sigma_{max}/\sigma_0 \approx 75$

Example 1.1b—creep failure with harmonic strain: $\varepsilon = \varepsilon_0(\sin(pt))$

If $E \neq$ const. $[E = (0.625 - 0.375*\tanh(c*(\theta - \theta_g)))]$; $\sigma_{max}/\sigma_0 \approx 75$ and n = const. then Equation (1.3) has numerical solution (using POLYMATH software [55]):

Data: $E_0 = 0.583(10^4)$ MPa; $\varepsilon_0 = 0.3(10^{-4})$; $c = 5$; $\theta_g = 4$

Note: If E = const. and n = const. then Equation (1.2) has an analytical solution:

$$\frac{d\sigma}{dt} = -\frac{1}{n}\sigma + \frac{E}{2n}\varepsilon_0(\sin pt) + E(\varepsilon_0 p)(\cos pt)$$

$$\sigma = A\sin(pt) + B\cos(pt)$$

$$Apn\cos(pt) - Bpn\sin(pt) + A\sin(pt) + B\cos(pt) =$$

$$= \varepsilon_0 E(0.5\sin pt + pn\cos pt)$$

$$\begin{cases} Apn + B = \varepsilon_0 Epn \\ A - Bpn = 0.5\varepsilon_0 E \end{cases} \quad (1.39)$$

$$A = \frac{\varepsilon_0 E[p^2n^2 + 0.5]}{[p^2n^2 + 1]}; B = \frac{\varepsilon_0 E[0.5p^2n^2 + pn]}{[p^2n^2 + 1]}$$

$$\tan\varphi = \frac{B}{A} = \frac{[0.5p^2n^2 + pn]}{[p^2n^2 + 0.5]}; \quad \sigma_{max} = \sqrt{A^2 + B^2}$$

It should be noted that the phase shift angle between strains and stresses in this case depends not only on the frequency of steady-state oscillations, but also on the ratio of the viscous and elastic components of this model. Thus, for example, the maximum value of the phase shift angle at n = 0.3 is at a frequency p = 10 Hz, while at n = 0.1 it is at a frequency p = 3 Hz. It is also interesting to note that when p tends to infinity, the value of the phase shift angle decreases monotonically and asymptotically approaches $\varphi = 0.5$, regardless of the value of n below).

n = 0.1 p = 10; n = 0.05 p = 20; n = 0.3 p = 3

Maximum stress σ_{max} vs. frequency of steady-state oscillations p is presented on Fig. 1.6.

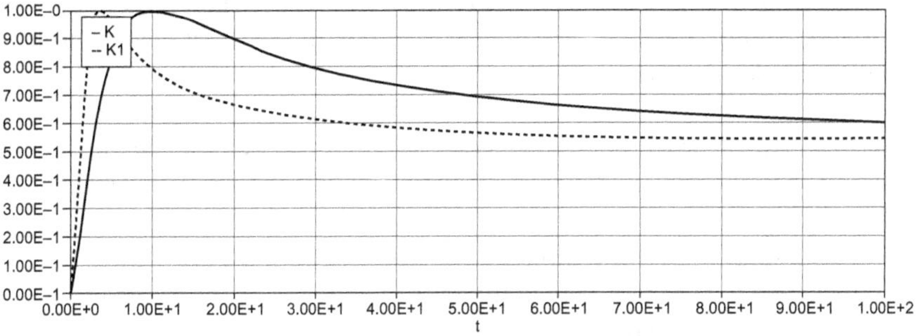

Figure 1.6. K = tan (φ) @ n = 0.3;'K1 = tan (φ) @ n = 0.1 t = p.

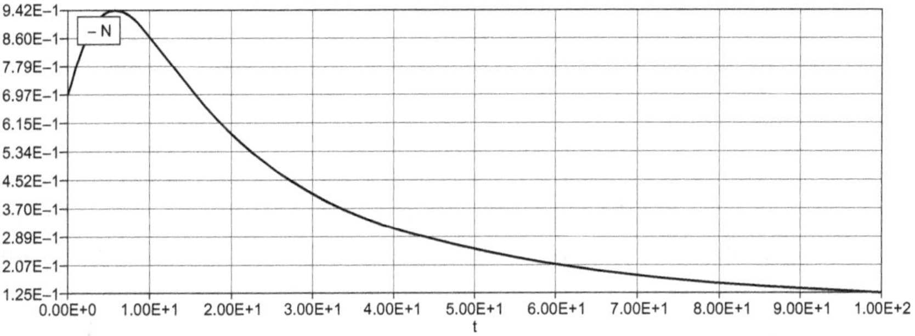

Figure 1.7. Maximum stress σ_{max} vs. frequency of steady-state oscillations p.

Obviously, the actual (dimensional) maximum stress has to be multiplied by $E\varepsilon_0$.

Example 1.2—If E ≠ const. [E = (0.625 – 0.375*tanh(c* (θ – θ_g)))] and n = const. then Equation (1.2) has numerical solution (using POLYMATH software [56]):

Data: E_0 = 0.583(10^4) MPa; ε_0 = 0.3(10^{-4}); c =5; θ_g = 4

Differential equations

1 d(Y)/d(t) = –((1/n)*Y*m1 + 0.29*2.01*e0*p*(cos (p*t))*E + 0.29*2.01*0.5*E*(1/n)*m1*(e0*(sin(p*t))))

Explicit equations

1 m = (0.0405*t – 0.01126*t^2 + 0.001462*t^3 – 0.00006868*t^4)
2 m1 = (0.0405 – 0.02252*t^1 + 0.004386*t^2 – 0.0002747*t^3)
3 E = (0.625 – 0.375*tanh(5* (t – 4)))
4 p = 1 Hz
5 n = 0.1
6 ε_0 = 0.3

For temperature range $0 < \theta < 11$ (300°F < T < 1260°F) the dynamic stress oscillations might be considered as periodical with the rapid change around transitional glass temperature T_g ($\theta_g = 4$).

The continuum mechanics of damage is based on the fact that the change in the mechanical properties of materials over time can be phenomenologically interpreted as the result of the accumulation of damage and various defects. When damage reaches a dangerous level, failure occurs. Cracking begins at the earliest stages of deformation and is associated with the growth of existing and the emergence of new sub- and micro defects. The material always has a large number of different defects, leading to high local stresses. The decrease in strength (the properties of the body to resist impacts from the external environment) of deformable solids can often be explained by the hidden destruction and micro defect structure of the body. Thus, since damage to a body significantly affects the nature of its destruction, it becomes obvious that both the fracture mechanics and the damage mechanics are designed to solve the main applied problem of estimating the safety margin of a solid [20].

The modeling of damage processes using the damage parameter ω, the scalar parameter of Kachanov – Rabotnov, which [22] was associated with the porosity of the material and it was assumed that the damage accumulation was due to the combined effect of diffusion and viscous pore growth mechanisms under conditions of high temperature creep. In [23], the intensity of accumulated creep strain is taken as a measure of material damage. The model proposed by the author within the framework of the "Standard linear mechanical system" of Voigt – Kelvin describes the cumulative damage under creep conditions.. The rate of cumulative damage ω is included in the dimensionless differential creep-fatigue equation of the Voigt – Kelvin model that has the following form,

$$\frac{d\sigma}{dt} = -\frac{1}{n}\sigma + \frac{E}{2n}\varepsilon_0(\sin pt)\left(\frac{1}{1-X}\right)^q + E(\varepsilon_0 p)(\cos pt)\left(\frac{1}{1-X}\right)^q$$

$$\frac{dX}{dt} = A\left(\frac{\sigma^l}{1-X}\right)^q; \quad A = \text{const.}; \quad t = m(T) \qquad (1.40)$$

$$E = E_1 = E_2; H = \frac{E_1E_2}{E_1 + E_2}; n = \frac{\eta}{E_1 + E_2}$$

$$\sigma(0) = 0; \quad X(0) = 0$$

The parameters 'q' and 'l' are obtained from the experimental cyclic fatigue data. Consider first the effect of temperature – time curves (linear and nonlinear) on the cyclic creep-fatigue process.

Example 1.3—Linear Temperature – time function: T = 10*t and t = 0.1T

For a small value "**φ**" (volumetric percentage of filler in a given composite structural element) the modulus of elasticity approximately equals, E_m – modulus of elasticity of the matrix material. Assuming that $E = E_1 = E_2 = E_m$ and n ≠ constant, we have:

Data: [E = (0.625 – 0.375*tanh(c* (θ – θ_g)))], then Equation (1.2) has numerical solution (using POLYMATH software): (see Example 1.2 above)

Differential equations

1 d(Y)/d(t) = –(1/n)*Y*m21 + 0.29*2.01*e0*(p^1)*(cos (p*t)^1)*E*(1/(1–X)^6) + (1/(1–X)^6)*0.29*2.01*0.5*E*(1/n)*(e0*(sin(p*t)^1))*m21

2 d(X)/d(t) = m21*A*(Y^2)/((1-X)^6)

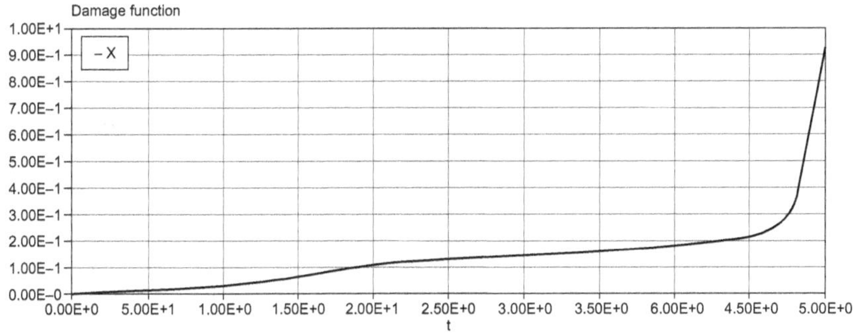

Figure 1.8. Damage diagram.

Model: X = a1*t + a2*t^2 + a3*t^3 + a4*t^4 + a5*t^5

Value
0.1674214
–0.3160684
0.239625
–0.0689635
0.0066824

$$X = \omega = 0.167\theta - 0.316\theta^2 + 0.24\,\theta^3 - 6.9(10^{-2})\,\theta^4 + 6.68(10^{-3})\,\theta^5 \qquad (1.41)$$

Example 1.4—Nonlinear Temperature – time function: T = m^{-1}(t) and t = m (T)

Calculated values of DEQ variables

	Variable	Initial value	Minimal value	Maximal value	Final value
1	A	2.	2.	2.	2.
2	a0	4.5	4.5	4.5	4.5
3	E	0.95	0.8824166	0.95	0.8824166
4	fi	0.05	0.05	0.05	0.05
5	k	0	0	0	0
6	m	0	0	0.0946124	0.0946124
7	m1	0.0405	0.0106296	0.0405	0.0149574
8	n	0.1	0.010544	0.1	0.010544
9	p	100.	100.	100.	100.
10	t	0	0	5.8	**5.8**
11	t1	6.	6.	6.	6.
12	t2	6.	6.	6.	6.
13	X	0	0	0.9171273	**0.9171273**
14	Y	0	-15.4538	1.705135	–15.4538

Differential equations

1 d(Y)/d(t) = –(1/n)*Y*m1 + 0.175*a0*(p)*(cos (p*t))*E*(1/(1–X)^6) + (1/(1–X)^6)*0.175*0.5*E*(1/n)*m1*(a0*(sin(p*t)))

2 d(X)/d(t) = m1*A*(Y^2)/((1–X)^6)

Explicit equations

1 m = (0.0405*t – 0.01026*t^2 + 0.001462*t^3 – 0.00006868*t^4)

2 m1 = (0.0405 – 0.02052*t^1 + 0.004386*t^2 – 0.0002747*t^3)

3 p = (100)

4 fi = 0.05

5 n = 0.10*(exp(–0.04*10*t))

6 a_0 = 4.5

7 A = 2

8 t1 = 6

9 E = (0.625 – 0.375*tanh(3* (t – t1)))

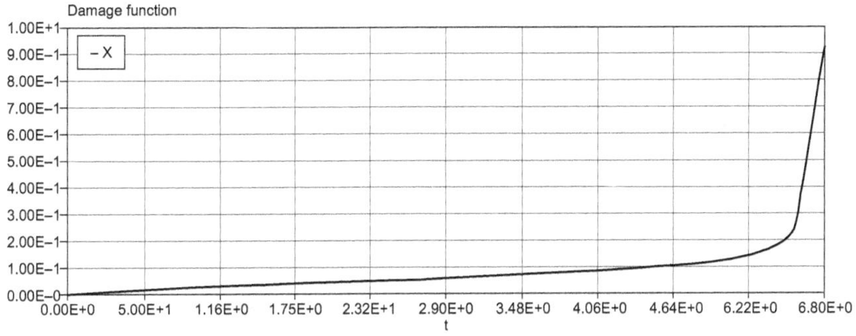

Figure 1.9. Damage diagram.

Model: X = a1*t + a2*t^2 + a3*t^3 + a4*t^4 + a5*t^5

Variable	Value
a1	0.1884348
a2	–0.3116415
a3	0.1840249
a4	–0.0432204
a5	0.0035065

$$X = \omega = 0.188\theta - 0.311\theta^2 + 0.184\,\theta^3 - 4.32(10^{-2})\,\theta^4 + 3.51(10^{-3})\,\theta^5 \qquad (1.42)$$

Comparing the results of Examples 1.3 and 1.4 it can be concluded that the maximum dimensionless temperatures θ_{max} are approximately the same (θ_{max} = 5 and θ_{max} = 5.8). Therefore, the total number of cycles before the creep-fatigue failure is almost the same. This type of simplification will be used in the creep-fatigue behavior analysis.

1.16a Standard Linear Model with different viscosity – temperature relationships

Exponential Model of viscosity. An exponential model for the temperature dependence of shear viscosity (μ) was first proposed by Reynolds in 1886.

$$\mu(T) = \mu_0 \exp(-bT) \qquad (1.43)$$

T is the temperature, μ and μ_0 μ 0 {\displaystyle \mu_{0}} and b {\displaystyle b} b are coefficients. This is an empirical model that usually works for a limited range of temperatures.

Arrhenius model

This model is based on the assumption that the fluid flow obeys the Arrhenius equation for molecular kinetics:

$$\mu(T) = \mu_0 \exp(\frac{E}{RT}) \tag{1.44}$$

T is temperature, μ_0 is a coefficient, E is the activation energy and *R* is the universal gas constant. A first-order fluid is another name for a power-law fluid with exponential dependence of viscosity on temperature.

Williams-Landel-Ferry model

The **Williams-Landel-Ferry** model, or **WLF** for short, is usually used for polymer melts or other fluids that have a glass transition temperature. The model is as follows:

$$\mu(T) = \mu_0 \exp\left(\frac{C_1(T - T_g)}{C_2 + T - T_g}\right) \tag{1.45}$$

T-temperature, C 1 {\displaystyle C_{1}}C_1, C_2 C 2 {\displaystyle C_{2}}, T_g T r {\displaystyle T_{r}} and μ 0 {\displaystyle \mu _{0}} are empirical parameters (only three of them are independent from each other). If one selects T_g (the parameter T r {\displaystyle T_{r}} based on the glass transition temperature), then the parameters C 1 {\displaystyle C_{1}} C_1,C_2 C 2 {\displaystyle C_{2}} become very similar for a wide class of polymers. Typically, we get C 1 ≈ {\displaystyle C_{1}\approx } C_1 =17.44K and C_2 = 51.6K.

Solid systems, unlike liquids, show signs of flow only after application of some limiting temperature. This means that their yield strength is not zero; their rheological curves do not pass through the origin, but are shifted from it by the amount of yield strength. The deformations (flow) of plastic and pseudo plastic solid systems, as well as the flow of liquids, are irreversible.

1.17 Temperature effect on viscosity

Typical relationship for the variation of viscosity η_0 with temperature would be:

$$\eta = \eta_0 [\exp \alpha(T - T_*)] \tag{1.46}$$

where η and T are the instantaneous viscosity and temperature respectively and η_0 is the viscosity at temperature T_*. For a polyester resin, a typical value of α would be –0.04°C.

Table 1.2a. Property comparisons for different types of materials (typical value and range).

Property	**Metals**	**Ceramics**	**Polymers**
Density ρ[kg/m3]	8000 (2000..22000)	4000 (2000..18000)	1000 (900..2000)
Thermal expansion α [1/K]	10·10-6(1·10-6.. 100·10-6)	10·10-6(1·10-6.. 20·10-6)	100·10-6(50·10-6.. 500·10-6)
Thermal capacity c_p [J/(kg·K)]	500 (100..1000)	900 (500..1000)	1500 (1000..3000)
Thermal conductivity k [W/(m·K)]	100 (10..500)	1 (0.1..20)	1 (0.1.. 20)
Melting (or yield) point T_m [K]	1000 (250..3700)	2000 (1000..4000)	400 (350..600)
Elastic Young's modulus E [GPa]	200 (20..400)	200 (100..500)	1 (10-3..10)
Poisson's ratio	0.3 (0.25..0.35)	0.25 (0.2..0.3)	0.4 (0.3..0.5)
Break strength σ_{break} [MPa]	500 (100..2500)	100 (10..400 tensile) (50..5000 compr.)	50 (10..150 tensile) (10..350 compr.)
Thermal creep resistance	Poor to Medium	Excellent	Very Poor
Hardness	Medium	High	Low
Thermal shock resistance	Good	Poor	Very Poor

The influence of temperature on the viscosity for Newtonian fluids can be expressed in terms of an Arrhenius type equation involving the absolute temperature (T)and the energy of activation viscosity (E_a) [24].

$$\mu = f(T) = A\exp(\frac{E_a}{RT}) \tag{1.47}$$

E_a and A are determined from experimental data. Higher E_a values indicate a more rapid change in viscosity with temperature. Considering an unknown viscosity μ at any temperature (T) and a reference viscosity μ_r at a reference temperature (T_r), the constant (A) may be estimated from Equation (1.27) and the resulting equation written in logarithmic form is:

$$\ln(\frac{\mu}{\mu_r}) = \frac{E_a}{R}(\frac{1}{T} - \frac{1}{T_r}) \tag{1.48}$$

In addition to modeling the viscosity of Newtonian fluids, the Arrhenius law can be used to model the influence of temperature on apparent viscosity in power law fluids. Considering a constant shear strain rate, with the assumption that temperature has negligible effect on the flow behavior index, yields

$$\ln(\frac{\eta}{\eta_r}) = \frac{E_a}{R}(\frac{1}{T} - \frac{1}{T_r}) \tag{1.49}$$

Or

$$\frac{\eta}{\eta_r} = \exp\left[\frac{E_a}{R}(\frac{1}{T} - \frac{1}{T_r})\right] \quad (1.50)$$

This equation can be used to find η at any temperature (T) from appropriate reference values (η_r, T_r).

The effect of shear strain rate and temperature can be combined into a single expression.

$$\eta = f(T, \dot{\gamma}) = K_T \exp(\frac{E_a}{RT})](\dot{\gamma})^{\bar{n}-1} \quad (1.51)$$

where n is an average value of the flow behavior index based on all temperatures. Equation (1.20) can also be expressed in terms of shear stress:

$$\sigma = f(T, \dot{\gamma}) = K_T \exp(\frac{E_a}{RT})](\dot{\gamma})^{\bar{n}} \quad (1.52)$$

The model (Equation 1.21), however, can be useful in solving many engineering problems such as creep deformation of composite materials.

Effect of temperature and concentration (C) of fillers in composite materials on apparent viscosity, at a constant shear strain rate, can also be combined into a single relationship.

$$\eta = f(T, C) = K_T [\exp(\frac{E_a}{RT})](\gamma) C^B \quad (1.53)$$

The three constants ($K_{T,C}$; E_a; B) must be determined from experimental data. Shear strain rate, temperature, and concentration (additives or moisture content) can also be combined into a single expression.

$$\eta = f(T, \dot{\gamma}, C) = K_{T,\dot{\gamma},C} [\exp(\frac{E_a}{RT})(\gamma) + B(C)] \quad (1.54)$$

Here the influence of shear strain rate is given in terms of a power law function.

1.18 Viscosity of dispersed systems

The viscosity of free-dispersed systems increases with increasing concentration of the dispersed filler. The presence of particles of the dispersed phase leads to a distortion of the fluid flow near these particles, which affects the viscosity of the dispersed system. If the concentration is insignificant, then the nature of the movement of the fluid around one of the particles will not affect the movement of the fluid near the other. Under these conditions, to determine the viscosity of free-dispersed systems, one can use the Einstein formula (flow) of plastic and pseudo plastic solid systems, as well as the flow of liquids.

$$\frac{\eta}{\eta_0} = 1 + s\varphi_f \qquad (1.55)$$

$$s = 2.5$$

η, η_0 is the viscosity coefficient of the free-dispersed system and the dispersion medium; s – coefficient depending on the shape of particles, for spherical particles s = 2.5; φ_f is the volume concentration of the dispersed phase. According to Einstein's formula, the viscosity of the solution does not depend on the size of the spherical particles . Subsequently, for particles having the shape of an ellipsoid, disk, dumbbell, and other three-dimensional particles the numerical value of the coefficient s changes. Einstein's formula is valid in the absence of particle deformation, if the concentration of the dispersed phase does not exceed 5%. When increasing the volume concentration of spherical particles up to 50% under conditions of mutual collision of particles, the formula to determine the viscosity is as follows:

$$\frac{\eta}{\eta_0} = 1 + 2.5\varphi_f + 14.7\varphi_f^2 \qquad (1.56)$$

The differential equation (1.4) in this case is as follows:

$$\frac{dY}{dt} = -\frac{Y}{n}(ml) + 0.175(p)(\cos(pt))E_0\left(\frac{1}{(1-X)^6}\right) +$$

$$+0.175\left(\frac{1}{(1-X)^6}\right)0.5E_0\frac{1}{n}(ml)(\sin(pt)) \qquad (1.57)$$

$$\frac{dX}{dt} = A(ml)\left(\frac{Y}{(1-X)^6}\right)$$

$$n = \frac{\eta(T)}{E_0} = \frac{\eta_0 \exp(-b(T-T_*))}{E_0}(1 + 2.5\varphi_f + 14.7\varphi_f^2)$$

Consider also the temperature effect and volumetric concentration of fillers on the modulus of elasticity, E, and viscosity parameter n (both of them as a function of dimensionless temperature θ and volumetric concentration

fi $\equiv \varphi$ – see Equation 1.2 (above) on cyclic creep-fatigue process behavior.

Example 1.5—Data – see Example 1.1, but $k = E_f/E_0 = 10$ and n = 0.1*(exp0 .04*10*t))*(1+2.5*fi+14.7*(fi^2))

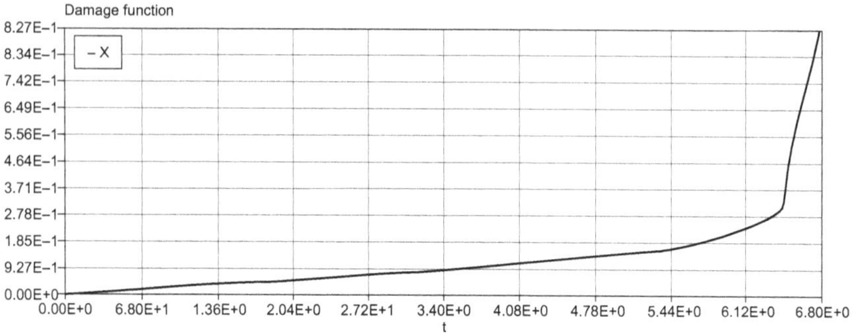

Figure 1.10. Damage function.

Model: X = a1*t + a2*t^2 + a3*t^3 + a4*t^4 + a5*t^5

Variable	Value
a1	0.1550189
a2	–0.2093686
a3	0.1070177
a4	–0.0216114
a5	0.0015076

$$X = \omega = 0.155\theta - 0.209\theta^2 + 0.107\theta^3 - 2.16(10^{-2})\,\theta^4 + 1.5(10^{-3})\,\theta^5 \tag{1.58}$$

The results and analyses of the above examples allow to draw several important conclusions that will be applied in the future presentation of more complex (integral form) models of cyclic creep-fatigue of composites and nanocomposites. In particular:

1. As a criterion for fatigue (durability) of a composite or nanocomposite, one can take its state in which the function of the total damage accumulation $\omega(\theta)$ takes the value $\omega(\theta) = 1$. Moreover, this function is a monotonously increasing function over the entire domain of increasing dimensionless temperature ($0 < \theta < \theta_{max}$).
2. It is obvious that the maximum value of the dimensionless temperature θ_{max} represents another criterion for the cyclic fatigue failure of the composite. It should be noted that the maximum value of the cycles to fatigue failure N_f is linearly related to the maximum value of the dimensionless temperature and the oscillation frequency "p", namely

$N_f = pt / 2\pi$. In turn, as mentioned above, the assumed external temperature-time ratio is given. Finally,

$N_f = pm / 2\pi$, where $t = m(\theta_{max})$. For instance, from Example 1.3: $m = 0.1\theta_{max}$; $p = 100$, therefore:

$N_f = p(0.1\theta_{max})/ 2\pi = 100(0.1)5.0 / 2\pi = 7.96$

3. Comparing results from Examples 1.4 and 1.3, one can see that the value of N_f is the same. However, the temperature – time functions are not the same: the latter is not linear, and in Example 1.3 has a linear relationship. Given the complexity of the formulation of the problem of cyclic creep - fatigue of composites, any simplification (not distorting the formulation of the problem as a whole) is the key to a successful approximate solution of the fatigue problem of composites and nanocomposites. Therefore, in the further research analyses, adapting the linear temperature-time function preferred.
4. In practice, any formulation and solution of the problem (in the engineering sense) of cyclic creep-fatigue of composites is possible only by introducing the so-called averaged (effective) parameters (such as the effective modulus of elasticity or the effective viscosity coefficient), and also effective dependent and independent variables (stress and strain, respectively). It should be noted that the high temperature loading has a very significant effect on both of these parameters E_{eff} and η_{eff}, as well as on the entire process of cyclic creep-fatigue of composites and nanocomposites. However, before proceeding to the analysis of the influence of these (and other) parameters on the cyclic creep-fatigue process of composites in the most general (integral) case, it makes sense to analyze these effects in the case of the simple linear Voigt – Kelvin model. On the other hand, the basic laws obtained by applying the simple Voigt – Kelvin model should coincide (at least from the qualitative point of view) with the results obtained in the general (integral) form of the problem.
5. Be noted that the damage function is monotonically increasing and having a upward concave curvature—concave up. In addition, it is important to note that this function increases sharply by the time the composite fails. Many experimental data confirm this. At the same time, the number of cycles until complete fatigue failure changes insignificantly, and for a better analytical approximation of the damage curve, a slightly smaller interval of dimensionless temperatures (of the order of 90% – 95%) is often used.

6. The introduction of the dimensionless parameters and dependent and independent variables allows the reduction of their number to a minimum and thus—the total number of experiments required for obtaining the internal parameters of the composite material. Some parameters are constant , and some are functions of dimensionless temperature (for example, modulus of elasticity or viscosity). The ratio between these parameters fundamentally affects the result—the cyclic fatigue of the composite or nanocomposite. Therefore, the analysis of these dependences and their influence on the fatigue of composites (including the construction of fatigue curves by numerical methods) is the focus of this chapter.
7. The model of cumulative damage of a composite material is based on the hypothesis that a material is considered a failure if the damage function is equal to unity at the maximum value of the dimensionless temperature.

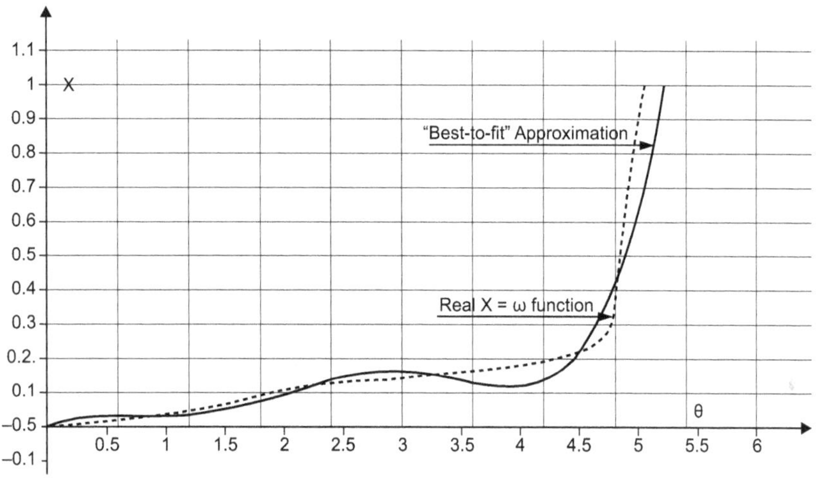

Figure 1.11. Damage v.s. dimensionless temperature diagram.

The most important tasks here are related to determining the conditions for the intensive development of cyclic creep, evaluating the interaction of fatigue and creep, as well as obtaining quantitative relationships between strain, cyclic creep rate, temperature and stress of the cycle. Of considerable interest are the problems of analytic estimation of so-called S – N fatigue curves and the limiting state of materials under conditions of cyclic creep. To describe the experimentally obtained high-temperature curves of creep and failure defining equations like Bailey-Norton (the time-hardening formulation of power law creep) and Rabotnov-Kachanov were used:

$$\dot{\varepsilon} = B\frac{\sigma^n}{(1-\omega)^m}; \quad \dot{\omega} = D\frac{\sigma^n}{(1-\omega)^m} \qquad (1.59)$$
$$\omega(0) = 0; \quad \omega(t^*) = \omega_2$$

Consider first the Reynolds formula (exponential relationship between viscosity and differential temperature).

Case 1: $\eta = \exp(-b(T - T_*)$ – Reynolds formula; $b = 0.04$

Differential Equation for Standard Linear Model:

$$\frac{dY}{dt} = -\frac{Y}{n}(m1) + 0.175(p)(\cos(pm))E_0\left(\frac{1}{(1-X)^6}\right) +$$
$$+0.175\left(\frac{1}{(1-X)^6}\right)0.5E_0\frac{1}{n}(m1)(\sin(pm))$$
$$\frac{dX}{dt} = A(m1)\left(\frac{Y}{(1-X)^6}\right) \quad X = \omega \qquad (1.60)$$
$$n = \frac{\eta_0[\eta(T)]}{E_0} = \frac{\eta_0[\exp(-b(T - T_*))]}{E_0}; \ \varepsilon = a_0\sin(pt); \ \sigma = Y$$
$$m = \left(0.0405t - 0.01026t^2 + 0.001462t^3 - 0.00006868t^4\right)$$
$$m1 = \left(0.0405 - 0.02052t + 0.004386t^2 - 0.0002747t^3\right)$$

Example 1.6—Data – see Example 1.5 but $a_0 = 4.5$

Differential equations

1 d(Y)/d(t) = –(1/n)*Y*m21 + 0.175*a0*(p)*(cos (p*m2))*E*(1/(1–X)^6)+ (1/(1–X)^6)*0.175*0.5*E*(1/n)*m21*(a0*(sin(p*m2)))

2 d(X)/d(t) = m21*A*(Y^2)/((1–X)^6)

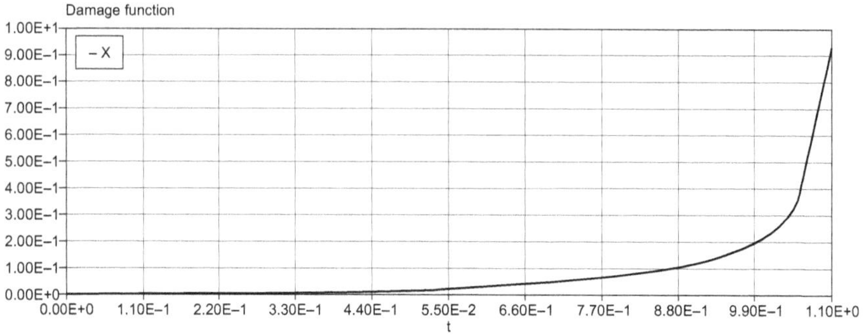

Figure 1.12. Damage function.

Model: X = a1*t + a2*t^2 + a3*t^3 + a4*t^4 + a5*t^5

Variable	Value
a1	0.7736081
a2	–7.704082
a3	24.54903
a4	–30.63665
a5	13.27239

$$X = \omega = 0.774\theta - 7.7\theta^2 + 25.55\theta^3 - 30.64\theta^4 + 13.27\theta^5 \quad (1.61)$$

$N_f = p(m2)/2\boldsymbol{\pi} = 1(0.11)/2 = 0.016$; $T_f = \boldsymbol{\beta} T_* + T_* = 0.1(600)\, \boldsymbol{\theta}_{max} + 600 = 666[^0K] \approx 366[^0C]$

Example 1.7—Data – see Example 1.6 but $a_0 = 4.0$

Differential equations

1 d(Y)/d(t) = –(1/n)*Y*m21 + 0.175*a0*(p)*(cos (p*m2))*E*(1/(1–X)^6) + (1/(1–X)^6)*0.175*0.5*E*(1/n)*m21*(a0*(sin(p*m2)))

2 d(X)/d(t) = m21*A*(Y^2)/((1–X)^6)

Damage Function

Model: X = a1*t + a2*t^2 + a3*t^3 + a4*t^4 + a5*t^5

Variable	Value
a1	0.616976
a2	–5.347539
a3	14.8589
a4	–16.11577
a5	6.070011

$$X = \omega = 0.617\theta - 5.35\theta^2 + 14.859\theta^3 - 16.116\theta^4 + 6.07\theta^5 \quad (1.62)$$

$N_f = p(m2)/2\boldsymbol{\pi} = 1(0.112)/2\boldsymbol{\pi} = 0.0179$; $T_f = \boldsymbol{\beta} T_* + T_* = 0.1(600)\ \boldsymbol{\theta}_{max} + 600 = 675[^0K] \approx 375[^0C]$

Repeating similar computer calculations for discrete values of a_0 from the interval

[1.0, 1.5, ..., 4.5], the results are as follows (see Table 1.3):

p = 1

Table 1.3. S – N_f data.

a_0	1.0	1.5	2.0	2.5	3.0	3.5	4.0	4.5
θ_{max}	θ=10 m2=0.1θ m21=0.1 **N_f = 0.16**	θ=8.0 **N_f = 0.127**	θ=5.5 **N_f = 0.0875**	θ=3.5 **N_f = 0.0557**	θ=2.55 **N_f = 0.0406**	θ=1.55 **N_f = 0.0247**	θ=1.1 **N_f = 0.0179**	θ=1.1 **N_f = 0.016**

Model: x = a0 + a1*y + a2*y^2

Variable	Value
a0	4.808375
a1	–46.55485
a2	147.8445

$S = a_0 = 4.8 - 46.55[N_f] + 147.84[N_f]^2$ a_0 – dimensionless cyclic stress amplitude

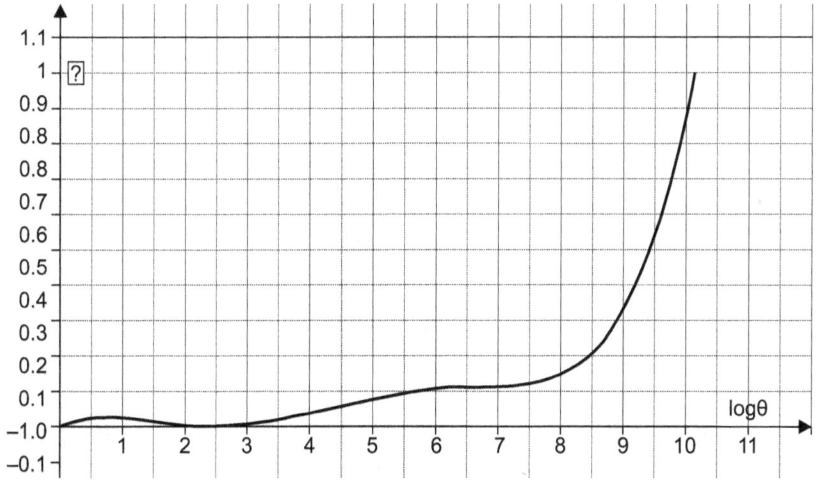

Figure 1.13. Relationship ω – θ (p = 1).

$$Y = \boldsymbol{\sigma} = 0.163\theta - 0.0834\theta^2 + 0.017\theta^3 - 0.00153\theta^4 + 5.052(10^{-5})\,\theta^5 \qquad (1.63)$$

Repeating similar computer calculations for discrete values of p from the interval [1 – 10^7] the results are as follows (see Table 1.4):

Table 1.4. Summary for a_0 [1 – 4.5] and p [1 – 10^7].

a_0	1.0	1.5	2.0	2.5	3.0	3.5	4.0	4.5
X=ω **p = 1**	2.856E-05	6.227E-05	2.227E-04	8.320E-04	1.523E-03	0.0514	0.08818	0.922 @ θ=39.8 $\mathbf{N_f = 0.63}$
X=ω **p = 10¹**	0.000725	0.00164	0.00294	0.00463	0.00675	0.00931	0.0124	0.922 @ θ=35.8 $\mathbf{N_f = 5.7}$
X=ω **p = 10²**	0.00287	0.00668	0.0125	0.021	0.0334	0.0531	0.0904	0.933 @ θ=2.0 $\mathbf{N_f = 3.18}$
X=ω **p = 10³**	0.00279	0.00648	0.0121	0.0203	0.0324	0.0514	0.08818	**0.922** @ θ=1.8 $\mathbf{N_f = 28.6}$
X=ω **p = 10⁴**	0.00334	0.00769	0.0141	0.0232	0.0359	0.055	0.0865	**0.9** @ θ=1.8 $\mathbf{N_f = 286.5}$
X=ω **p = 10⁵**	0.00277	0.00644	0.0120	0.0201	0.0321	0.0508	0.0866	**0.911** @ θ=1.779 $\mathbf{N_f = 2{,}865}$
X=ω **p = 10⁶**	0.00277	0.00644	0.0120	0.0201	0.0321	0.0508	0.0866	**0.911** @ θ=1.779 $\mathbf{N_f = 2{,}865}$
X=ω p = 10⁷	0.00277	0.00644	0.0120	0.0201	0.0321	0.0508	0.0866	**0.911** @ θ=1.779 $\mathbf{N_f = 2{,}865}$

The maximum time before failure is (from value 'm'—see data results above): $t_{max} \approx 0.139(4)/0.2 = 2.75$ hours; and the maximum strain is: $\varepsilon = 7.02(10^{-4})11.25 = 0.0079$. It should be noted that the frequency of the material response is changing drastically around the glass transitional temperature $\theta_g = 5$ (T = 350°C) due to changes in the instantaneous modulus of elasticity. This effect and many others, such as: the combined effect of static and cyclic loads; filler's modulus of elasticity; functions of material damage and many others parameters and variables on creep-fatigue process are discussed in the corresponding chapters. It is a well-known fact that an increased number of standard elements connected in parallel (or sequentially), can be replaced by the Volterra integral equation of the second kind (with the number of standard elements tending to infinity). At the same time, to analyze the process of dynamic creep-fatigue of composite materials, it is necessary to consider three possible options, namely:

1. Temperature dependence of time (external load on the composite) is a monotonically increasing (or decreasing) function. The forced oscillations of the composite are caused by instantaneous relative axial elongations of the composite to a given sinusoidal law in time.
2. Temperature dependence of time (external load on the composite) is a given sinusoidal law in time. In this case, the instantaneous relative axial elongations of the composite are subject to the linear law of elongation (compression)—linear function of temperature.
3. Dependence of temperature on time (external load on the composite) is a given sinusoidal law in time. In this case, the instantaneous relative axial elongations of the composite are also governed by the given sinusoidal law in time.

It should be noted that in all three cases it is assumed that the effective modulus of elasticity of a composite obeys the well known 'mixture rule' and it is a function of dimensionless temperature θ only.

In principle, for each structure, a material can be developed that most fully corresponds to its purpose, field of acting loads and operating conditions. In this respect, composites are similar to natural materials, whose rational combination of properties evolved over a long course of time. The directional nature of the properties of composites simultaneously means that, along with high characteristics in some directions, they have lower characteristics in others. Therefore, making an arbitrarily bad construction from any good composite is much easier than making one with metal.

However, proper consideration of the peculiarities of composites makes it possible to obtain structures with a small degree of weight/strength ratio, an improvement that cannot be achieved using traditional materials.

References

[1] Van Paepegem W. Degrieck J. Fatigue Damage Modeling of Fiber-reinforced Composite Materials // Review, 2000, Applied Mechanics Reviews, 54(4): 279–300, 2000.

[2] Harris B. Fatigue in Composites // CRC Press, Boca Raton, pp. 10–68, 2003.

[3] Hashin Z. and Rotem A. A fatigue criterion for fiber reinforced composite materials // Journal of Composite Materials, 7: 448–464, 1973.

[4] Talreja R. Fatigue life modeling. Second International Conference on Fatigue of Composites, Williamsburg, VA, 2000.

[5] Hashin Z. Cumulative damage theory for composite materials: residual life and residual strength methods. Composite Science and Technology, 23: 1–19, 1985.

[6] Lemaitre J., ed. Handbook of Materials Behavior Models. San Diego, Academic Press, 2001.

[7] Lemaitre J. A Course on Damage Mechanics. Berlin: Springer, 1996.

[8] Kattan P. I. and Voyiadjis G. Z. Damage Mechanics with Finite Elements. Berlin, Springer, 2001.
[9] Razdolsky L. Phenomenological Creep Models of Composites and Nanomaterials Deterministic and Probabilistic Approach. CRC Press Taylor & Francis Group, Boca Raton, FL.
[10] Talreja R. Fatigue life modeling. Second International Conference on Fatigue of Composites, Williamsburg, VA, 2000.
[11] Talreja R. Stiffness properties of composite laminates with matrix cracking and interior delamination. Engineering Fracture Mechanics, 25(5/6): 751–762, 1986.
[12] Hayder Al-Shukri and Muhannad Khelifa. Fatigue study of E-glass fiber reinforced polyester composite under fully reversed loading and spectrum loading. Engineering & Technology, 26(10), 2008.
[13] Hader Al-Shukri.Experimental and theoretical investigation into some mechanical properties of E-glass polyester composite under static and dynamic loads. Engineering and Technology, 2007.
[14] Rita R. and Bose N. R. Behavior of E-glass fiber reinforced vinyl ester resin composites under fatigue condition. Bulletin Materials Science, 24(2): 137–142, 2001.
[15] Leo Razdolsky, AIAA – 2019, Fatigue – Creep Phenomenological Models of Composites and Nanocomposites, 2019.
[16] Philippidis T. P. and Vassilopoulos A. P. Fatigue design allowables of GRP laminates based on stiffness degradation measurements. Composite Science and Technology, pp. 2819–2828, 2000.
[17] Schutz, W. A history of fatigue. Engineering Fracture Mechanics, 54(2): 263–300, 1996.
[18] Kim D. W. and Chang J. H. Evaluation of the creep-fatigue damage mechanism of type 316L and type 316LN stainless steel. Int. J. Press. Vess. Piping, 85: 378–384, 2008.
[19] Dowling N. E. Mean stress effects in stress-life and strain-life fatigue. In: Proc. of the Third Int. Sea Fatigue Congress, Sao Paulo, Brazil, 2004.
[20] Boyle J. T. and Spence J. Stress Analysis for Creep, Butterworth & Co. Ltd., 1983.
[21] Van Paepegem W. and Degrieck J. Fatigue Damage Modeling of Fiber reinforced Composite Materials // Review, 2000, Applied Mechanics Reviews, 54(4): 279–300, 2000.
[22] Hashin Z. and Rotem A. A fatigue criterion for fiber reinforced composite materials // J Composite Mat., 7: 448–464, 1973.
[23] Djimedo Kondo, Hélène Welemane and Fabrice Cormery. Basic concepts and models in continuum damage mechanics. Revue europÃ©enne de gÃ©nie civil, 11: 927–943, 2007.
[24] Hashin Z. Cumulative damage theory for composite materials: residual life and residual strength methods // Composite Sci. Tech., 23: 1–19, 1985.
[25] Lemaitre J. ed. Handbook of Materials Behavior Models. San Diego: Academic Press, 2001.
[26] Lemaitre J. A Course on Damage Mechanics. Berlin: Springer, 1996.
[27] Kattan P. I. and Voyiadjis G. Z. Damage Mechanics With Finite Elements. Springer, 2001.
[28] Hayder Al-Shukri. Muhannad Khelifa Fatigue study of E-glass fiber reinforced polyester composite under fully reversed loading and spectrum loading // Eng. & Technology, 26(10), 2008.
[29] Endo, Tatsuo, Mitsunaga, Koichi, Takahashi, Kiyohum, Kobayashi, Kakuichi and Matsuishi, Masanori. Damage evaluation of metals for random or varying loading—three aspects of rain flow method. Mechanical Behavior of Materials, 1: 371–380, 1974.
[30] Sunder, R., Seetharam, S. A. and; Bhaskaran, T. A. 1984. Cycle counting for fatigue crack growth analysis. International Journal of Fatigue, 6(3): 147–156.
[31] Taira S. Lifetime of Structures Subjected to Varying Load and Temperature. Creep in Structures. Nicholas J. Hoff, ed., Academic Press, pp. 96–124, 1962.

[32] Kachanov L. M. Introduction to Continuum Damage Mechanics (Martinus Nijhoff Dortrecht, The Netherlands, 1986).
[33] Choy, Tuck C. 1999. Effective Medium Theory. Oxford: Clarendon Press.
[34] Wang M. and Pan, N. Predictions of effective physical properties of complex multiphase materials. Materials Science and Engineering: R: Reports. 63: 1–30, 2008.
[35] Landau L. D. and Lifshitz E. M. Electrodynamics of Continuous Media, Pergamon, 474 p., 1984.
[36] Lunkenheimer P. Dielectric spectro-scopy of glassy dynamics, Shaker Verlag, Aachen, 1999.
[37] Odquist F. and Hult J. Kriechfestigkeit metallischer Werkstoffe. Springer-Verlag, Berlin, 1962.
[38] Rabotnov Yu. N. Creep of structural elements. M .: Science, 752 p., 1966.
[39] Harris B. Fatigue in Composites. CRC Press, Boca Raton, pp. 10–68, 2003.
[40] Arrhenius S. A. 1889. Über die Reaktionsgeschwindigkeit bei der Inversion von Rohrzucker durch Säuren. Z. Phys. Chem. 4: 226–48.
[41] Hashin Z. Failure Criteria for Unidirectional Fiber Composites. Transactions of the ASME. Journal of Applied Mechanics, 45(2): 329–334, 1980.
[42] Puck W. Schneider. On failure mechanisms and failure criteria of filament-wound glass-fibre/resin composites. Plastics and Polymers February 1969. The Plastics Institute Transactions and Journal, Pergamon press, pp: 33–44, 1969.
[43] Wicaksono S. and Chai G. B. A review of advances in fatigue and life prediction of fiber-reinforced composites. Journal of Materials: Design and Applications, 227(3): 179–195, 2012.
[44] Degrieck J. and Paepegem W. V. Fatigue damage modeling of fiber-reinforced composite materials: Review. American Society of Mechanical Engineers, 54(4): 279–300, 2001.
[45] Halpin J. C., Jerina K. L. and Johnson T.A. Characterization of composites for the purpose of reliability evaluation. In Analysis of Test Methods for High Modulus Fibers and Composites; ASTM STP: West Conshohocken, PA, USA, pp. 5–64, 1973.
[46] Shokrieh M. M. and Lessard L. B. Multiaxial fatigue behavior of unidirectional plies based on uniaxial fatigue experiments - I. Modeling. International Journal of Fatigue, 19(3): 201–207, 1997.
[47] Manson S. S., Halford G. R. and Hirschberg M. H. Creep-fatigue analysis by strain range partitioning. In: Symposium on design for elevated temperature environment, ASME, New York, pp. 12–28, 1971.
[48] Kachanov L. M. Introduction to Continuum Damage Mechanics. Kluwer Academic Publ., Dordrecht, 1986.
[49] Lemaitre J. A Course on Damage Mechanics. Springer, Berlin, 1992.
[50] Razdolsky L. AIAA – 2017 Probability Based High Temperature Creep of Composites and Nanocomposites, 2017.
[51] Lemaitre J. A Course on Damage Mechanics. Springer, Berlin, 1992.
[52] Lemaitre, J. (edt.): Handbook of Materials Behaviour Models. Academic Press, San Diego, vol. 2, 2001.
[53] Razdolsky L. Fatigue-Creep Phenomenological Models of Composites and Nanocomposites. AIAA Propulsion and Energy Forum, AIAA, 2019.
[54] Lemaitre J. and Desmorat R. Engineering Damage Mechanics - Ductile, Creep, Fatigue and Brittle Failures. Springer, Berlin, 2007.
[55] Razdolsky L. Probability-Based Structural Fire Load. Cambridge University Press, UK, 2014.
[56] POLYMATH software.com.

CHAPTER 2

Cumulative Damage Model (CDM) of Cyclic Creep-Fatigue Process

2.1 Introduction

The simplest models considered in the previous chapter do not even qualitatively describe the basic viscoelastic properties of composites and nanocomposites. Purely elastic deformation is mechanically completely reversible and is not associated with creep. However, in a real life situation of viscoelastic composite solids, the energy and entropic elastic deformation is a viscous flow. This results in stress relaxation under constant deformation, creep under constant load, and energy dissipation under dynamic action (cyclic load). Therefore, when simulating the macroscopic mechanical properties of viscoelastic composite solids, even in the deformation region, elastic elements with damping should be used containing springs (modulus E) and elements that take into account losses depending on the rate of deformation (a damper characterized by viscosity μ). The high elasticity of structured composites formed by the interlacing of fibrous particles, as well as macromolecular chains , is primarily associated with the deformability of the fibers themselves and macromolecules. As is known, equations based on simple mechanical Kelvin – Voigt models (elastic and viscous elements connected in parallel) do not allow one to quantitatively describe the behavior of highly elastic structural systems. In modern literature, the description of the kinetics of elastic deformation and stress relaxation in such systems has become widespread using the concept of a spectrum of relaxation periods corresponding to a combination of a variety of elastic and viscous elements. At the same time, the kinetics of the development and decay of nonlinear viscoelastic deformation of a number of composite structured systems, as is well known, can be described by the Volterra integral equation of the second kind having the following form [1]:

$$E[t]\varepsilon = \sigma_0(t) + \int_0^t A_1 e^{-\frac{E_a}{RT}} K_1(t,\tau)\sigma^n(\tau)d\tau$$

$$\beta = \frac{RT_*}{E_a};\quad T = \beta T_*\theta + T_*;\quad T_* - \text{ Base temperature } [^0K] \qquad (2.1)$$

E_a – activation energy; R - universal gas temperature

σ_0 - instantaneous stress

The main assumption in the model formulation is that the total strain tensor can be expressed as the sum of instantaneous, inelastic, creep and thermal strains,

$$\varepsilon_{ij} = \varepsilon_{ij}^{inst} + \varepsilon_{ij}^{e} + \varepsilon_{ij}^{c} + \varepsilon_{ij}^{T} \qquad (2.1a)$$

ε_{ij} – component of total strain tensor, ε_{ij}^{inst} – component of instantaneous inelastic strain tensor,

ε_{ij}^{e} – component of inelastic (elastic) strain tensor, ε_{ij}^{c} – component of creep strain tensor,

ε_{ij}^{T} – component of thermal strain tensor.

This assumption allows the use of the so-called classical theories of creep which make a distinction between time-dependent and time-independent inelastic strains. The constitutive law for an isotropic, thermo-inelastic material with temperature-dependent modulus is [2, 3],

$$E[t]\varepsilon = \sigma_0(t) + \int_0^t A_1 e^{-\frac{E_a}{RT}} K_1(t,\tau)\sigma^n(\tau)d\tau \qquad (2.2)$$

$K_1(t,\tau)$ – creep function

Unlike metals, composite materials are substantially heterogeneous and anisotropic. Various methods of nondestructive testing can detect a change in the elastic modulus and damping parameters. The damageability can be accompanied by thermal effects and be recognized by appropriate thermal emission methods. Although the technology of composite materials is developing rapidly, their use in real structural design is hampered by the lack of available datasets on fatigue, which contain the main operational parameters in operation. Fatigue structures are made of composite materials—a very complex phenomenon. Analysis of the stress state at variable cyclic loads requires consideration of the anisotropy of the effective elastic and viscous properties of the composite. Accumulation of damage occurs not in a localized form, and the fatigue failure does not always occur because of the progression of one macroscopic crack. Micro structural damage accumulation mechanisms, including fiber or matrix fracture, fiber-matrix splitting and delaminating sometimes occur independently. At low levels of cyclic loading or in the initial part of the live term most types of composites accumulate scattered damage. These damages get distributed throughout the stress zone and gradually reduce the strength and stiffness of the composite. In the later stages of the life span of a composite, the accumulated damage in a certain area of the composite can be quite large. This leads to the fact that the residual strength of the composite area drops to minimum stress levels in cyclic loading that result in failure. Although the fatigue behavior of composite materials is significantly different from that of metals, many models have been developed based on the well-known S – N curves. These models make up the first class of the so-called "fatigue strength models". This approach requires large experimental studies and does not take into account real damage mechanisms such as matrix fractures and fiber breaks [4]. The second class includes phenomenological models for high cycle fatigue. These models offer an evolutionary law that describes gradual degradation of the strength or stiffness of the composite sample based on macroscopic properties.

Recently, the fatigue models based on concepts of continuum mechanics have been proposed [5]. In these models this type of damage quantitatively described by some internal parameters of the material. The development of damage is determined by evolutionary kinetic equations reflecting the irreversible nature of damage [6]. Continuum damage models introduce scalar, vector or tensor parameters. These models are based on physical modeling of the main damage mechanisms, which lead to macroscopically noticeable degradation of mechanical properties [7]. The main result of all the fatigue patterns is prediction of fatigue life and each of these categories uses its own criterion in order to determine the failure condition and, as a result, fatigue life of the composite material.

In this manuscript, we consider the continuum damage model based on the assumption that the growth of the damage function depends on the effective value of the cyclic creep stress of the composite and damage function. In this case, the cyclic creep-fatigue stress is a solution of the integral type of creep-fatigue equation, which includes the damage function ω (t, σ) that is similar to one described in [8–10]. Thus, the system of integro-differential equations has the following form [11,12]:

$$E[T(t)][a_0\sin(pt)] = \sigma(t) + \int_0^t e^{\frac{E_a}{RT}} \frac{\mathbf{1}}{(\mathbf{1}-\omega)^r} K_1(t,\tau)\sigma^n(\tau)d\tau +$$

$$+\int_0^t A_2 e^{\frac{E_a}{RT}} \frac{\mathbf{1}}{(\mathbf{1}-\omega)^r} f_2[\sigma(\tau)]K_2(t,\tau)\sigma^n(\tau)d\tau +$$

$$+\int_0^t \frac{E_0}{E_1} A_3 e^{\frac{E_a}{RT}} \frac{\mathbf{1}}{(\mathbf{1}-\omega)^r} f_3[d(\tau)]K_3(t,\tau)\sigma(\tau)d\tau + \sigma_{st}$$

$$\frac{d\omega}{dt} = A\frac{\sigma^q}{(\mathbf{1}-\omega)^r};\quad \omega(0) = \mathbf{0};\quad \sigma(\mathbf{0}) = \mathbf{0}$$

$$K_1(t,\tau) = \varphi_1(t)f_1(\tau) = \sum_{i=1}^{N}[\sin(t-\tau)]\exp(-\alpha_i t)\exp(\alpha_i\tau) \tag{2.3}$$

$$K_2(t,\tau) = \sum_{i=1}^{N}[\sin(t-\tau)]\exp(-\beta_{i2}t)\exp(\beta_{i2}\tau)$$

$$K_3(t,\tau) = \sum_{i=1}^{N}[\sin(t-\tau)]\exp(-\beta_{i3}t)\exp(\beta_{i3}\tau)$$

$$E = \left(a - b\left(\tanh\left(c\left(T - T_{gm}\right)\right)\right)\right)(\mathbf{1}-\varphi) + \left(a - b\left(\tanh\left(c\mathbf{1}\left(T - T_{gf}\right)\right)\right)\right)\varphi / k$$

$$f_2(\sigma) = \sigma^s;\quad s = \mathbf{1},\mathbf{2},\mathbf{3}\ldots, M;$$

$k = E_0/E_1;\ A_1 = \eta_0/E_0;\ \eta = \exp\left(-b\left(T - T_*\right)\right)$ – Reynolds formula

$A_2 = A_1(\eta)(((1+2.5\varphi+14.7(\varphi^2))d_{eff}$ – Einstein formula

$d = a_1 T;\ A_3 = [f_3(T)][d_{eff}(T)];\ d_{eff} = (\varphi d)/(1 + 0.333(1 - \varphi)d)$

2.2 The concept of effective stress

In the continuum approach to the analysis of the joint stress state of creep and fatigue of composites, the material is considered as a homogeneous anisotropic viscoelastic medium [13]. When building the model assumes that there are small deformations (from a geometric point of view).

The functional of the continuum damage rate is a positive non-decreasing function of time, and there is also a linear relationship between the effective stress tensor σ and the strain tensor ε for the complete composite package (for example, in the case of a multilayer system). Unlike brittle fracture mechanics, which considers the process of equilibrium or growth of macro cracks, continuum mechanics of damage uses continuous internal variables that are related to the density of micro defects. The proposed model uses the concept of effective stress and integrally reflects various types of damage at the micro scale level (from the standpoint of heredity and non-invariance in time of the creep process under cyclic loading). For example, the continuum damage evolution includes the formation and growth of clusters of nanoparticles; the presence of 2 (or more) phase states of the composite system; physicochemical effects associated with high-temperature changes; abrupt changes in stiffness (fiber breaks, delamination) and other microscopic defects [14]. The complexity and variety of mechanisms for the accumulation of fatigue damage and degradation of strength composite properties make justifiable use of scalar intrinsic variables for quantitative descriptions of damage [15].

From the perspective of the mechanics of the damaged environment the creep-fatigue process is associated with non-isothermal viscoelastic deformation and damage accumulation in structural composite materials in arbitrary non-stationary thermal load modes is modeled. A cumulative damage model consists of three interrelated components [16]:

- Relations determining the viscoelastic behavior of composite materials and their dependence on the damage function;
- Equations describing the kinetics of damage accumulation;
- Failure criteria of damaged composite materials.
- This version of the integral type constitutive equations of the creep-fatigue state considers the main inelastic deformation effects of composites and nanocomposites subjected to *monotonous and cyclic loads*. The development of the kinetic equations of damage accumulation is based on conservation of energy principles that take into account the main stages of the damage process: nonlinear summation of damage, influence of the type of deformation, parameters of stress (strain) limiting state and effect of accumulated damage to the processes of nucleation, growth and fusion of micro defects.

Physical and mechanical damage of a composite's characteristics are attributed to the influence of parameters such as: decrease in elastic modulus, fall stress amplitudes with constant strain amplitude and increase in strain amplitude with constant stress amplitude. An assessment of the adequacy of the developed models of the numerical studies of the viscoelastic deformation of cyclic processes and decomposition of composite materials maps the results obtained with the data from field experiments and the results obtained by other researchers. The mode of material testing for fatigue under the action of cyclic stresses, efforts of a given amplitude is so-called soft, and the test mode when the strain amplitude remains constant is hard. In tests with a given load the object of observation is deformations, displacements; when testing is conducted with a given deformation, a change in stresses is observed. To distinguish between soft and hard loading regimes is especially important in cases where the deformations are not completely elastic, for example, with low-cycle fatigue. When testing with a given deformation, the load value gradually decreases with an increasing number of cycles or when testing with a given load, the deformation value increases with an increase in the number of loading cycles, then the material is called cyclically weakening. If, with an increase in the number of loading cycles when testing with a given deformation, the load increases or the deformation gradually decreases then the material under test is called cyclically hardening. There are also so-called cyclically stable materials for which the load-strain diagram practically repeats from cycle to cycle, without changing quantitatively.

The phenomenon of hardening or softening under cyclic loading is peculiar not only for different materials, but also for different states of the same material. Obtaining the laws of cyclic deformation and under uniaxial and complex stress state experimentally is more complex than purely static experiments. In this connection, the problem arises off a theoretical description of cyclic creep curves up to fatigue failure using the results of experiments performed under static load. The solution of this problem for the case of high-cycle loading is given in [2] and for low-cycle loading in [3].

2.3 Classification of composite materials

One simple scheme for the classification of composite materials consists of two main divisions: particle-reinforced and fiber-reinforced and two subdivisions exist for each. The dispersed phase for particle-reinforced composites is equiaxed (i.e., particle dimensions are approximately the same in all directions and the media is considered as a homogeneous material); for fiber-reinforced composites, the dispersed phase has the geometry of a fiber (i.e., a large length-to-diameter ratio). The discussion of the remainder of this book will be organized according to this classification scheme. Large-particle and dispersion-strengthened composites are the two sub-classifications of particle-reinforced composites (see Fig. 2.1). The distinction between these is based upon reinforcement or strengthening mechanism. The term "large" is used to indicate that particle–matrix interactions cannot be treated on the atomic or molecular level; rather, continuum mechanics is used. For most of these composites, the large-particle phase is harder and stiffer than the matrix. These reinforcing particles tend to restrain movement of the matrix phase in the vicinity of each particle. In essence, the matrix transfers some of the applied stress to the particles, which bear a fraction of the load. The degree of reinforcement or improvement of mechanical behavior depends on strong bonding at the matrix–particle interface. For dispersion-strengthened composites, particles are normally much smaller, with diameters between 10 and 100 nm. Particle–matrix interactions that lead to strengthening occur on the atomic or molecular level. The matrix bears the major portion of an applied load; the small-dispersed particles hinder or impede the motion of dislocations. Thus, inelastic deformation is restricted such that yield and tensile strengths, as well as hardness, increase. Some polymeric materials to which fillers have been added are really large-particle composites. Again, the fillers modify or improve the properties of the material and/or replace some of the polymer volume with a less expensive material—the filler. Another familiar large-particle composite is concrete, which is composed of cement (the matrix), and sand and gravel (the particulates). Particles can have quite a large variety of geometries, but they should be of approximately the same dimensions in all directions (equiaxed). For effective reinforcement, the particles should be small and evenly distributed throughout the matrix. Furthermore, the volume fraction of the two phases influences the behavior; mechanical properties are enhanced with increasing particulate content. Two mathematical expressions have been formulated for the dependence of the elastic modulus on the volume fraction of the constituent phases for a two-phase composite.

Matrix material holds the imbedded phase and shares the load with the secondary phase

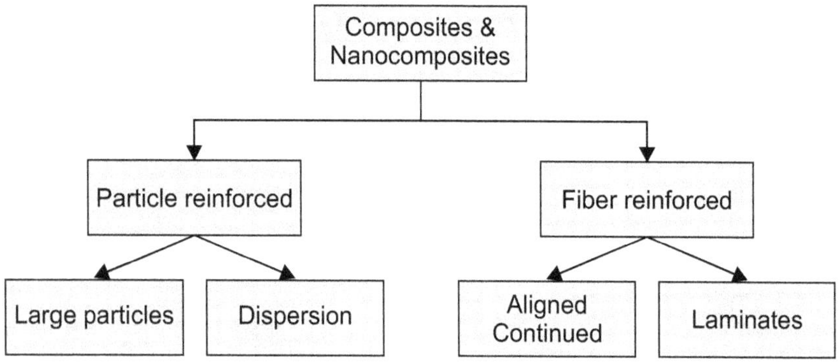

Metal alloys may be strengthened and hardened by the uniform dispersion of several volume percentages of fine particles of a very hard and inert material. The dispersed phase may be metallic or nonmetallic; oxide materials are often used. Again, the strengthening mechanism involves interactions between the particles and dislocations within the matrix, as with precipitation hardening. The dispersion strengthening effect is not as pronounced as with precipitation hardening; however, the strengthening is retained at elevated temperatures and for extended time periods because the dispersed particles are chosen to be nonreactive with the matrix phase. For precipitation-hardened alloys, the increase in strength may disappear upon heat treatment because of precipitate growth or dissolution of the precipitate phase. For precipitation-hardened alloys the "growth" of clusters follows the general rule of mixtures. In the two-dimensional case the transverse instantaneous modulus of elasticity shell be as follows:

$$(1/E_c) = V_m/E_m + V_f\,E_f \text{ or } E_c = E_m E_f/(V_m E_f + V_f E_m) \tag{2.1a}$$

Fiber-reinforced composites are sub classified by fiber length. For a short fiber, the fibers are too short to produce a significant improvement in strength. One-dimensional: maximum strength and stiffness are obtained in the direction of the fiber Planar: in the form of a two-dimensional fabric. Random or three-dimensional: the composite material tends to possess isotropic properties. This is the most widely used form if a laminar structure is made by stacking and bonding thin layers of fiber and polymer until the desired thickness is obtained. The mechanical characteristics of a fiber-reinforced composite depend not only on the properties of the fiber, but also on the degree to which an applied load is transmitted to the fibers by the matrix phase. The magnitude of the interfacial bond between the fiber and matrix phases is important to the extent of load transmittance. Under an applied stress, this fiber-matrix bond ceases at the fiber ends, in other words, there is no load transmittance from the matrix at each fiber extremity. A critical fiber length is necessary for effective strengthening and stiffening of the composite material. This critical length l_c is dependent on the fiber diameter d and its ultimate (or tensile) strength σ_f and on the fiber-matrix bond strength (or the shear yield strength of the matrix, whichever is smaller) and τ_c according to (2.2).

$$l_c = \frac{\sigma_f d}{2\tau_c} \tag{2.2a}$$

For a number of glass and carbon fiber-matrix combinations, this critical length ranges between 20 and 150 times the fiber diameter. When a stress equal to σ_f is applied to a fiber having just this critical length l_c, the maximum fiber load is achieved only at the axial center of the fiber. As fiber length increases to $l > l_c$, the fiber reinforcement becomes more effective; and the applied stress is equal to the fiber strength. Fibers for which $l >> l_c$ (normally $l > 15l_c$) are termed continuous; discontinuous or short fibers have lengths shorter than this. For discontinuous fibers with $l < l_c$, the matrix deforms around the fiber such that there is virtually no stress transference. To affect a significant improvement in the strength of the composite, the fibers must be continuous. The precipitates play the same role as particle substances in particle-reinforced composite materials. Just as the formation of ice in the air can produce clouds, snow, or hail, depending upon the thermal history of a given portion of the atmosphere, precipitation in solids can produce many different sizes of particles, which have radically different properties. Unlike ordinary tempering, alloys must be kept at elevated temperature for hours to allow precipitation to take place. This time delay is called aging. Solution treatment and aging is sometimes abbreviated "STA" in metal specs and certifications.

2.4 Objectives of this research

Materials can be hardened by inhibiting the motion of crystal defects called dislocations. In pure metals, the presence of defects (such as vacancies, interstitials, dislocations and grain boundaries) can enhance the strength. In single-phase alloys, additional resistance to deformation may arise from the presence of foreign atoms. In two-phase alloys, additional stress is needed to enable the dislocation to intersect the second-phase particles. The composite material can be strengthening by adding a finely dispersed precipitate. This phenomenon is termed precipitation hardening. The thermodynamics of precipitation in a composite can be best understood by referring to the binary phase diagram shown in Fig. 2.1. When the alloy of less than 5 wt percentage is heated to a temperature just above the solvus line, only one phase is thermodynamically stable. Other solid phases dissolve (disappear). This process is called solution treatment. The only requirement is that the specimen must be kept at this temperature for a sufficiently long time. When solution treated samples are rapidly cooled (quenched) to below the solvus line (Fig. 2.1), the two phases are thermodynamically stable. These phases are two different solids, physically distinct, and separated by a phase boundary. The process is similar to precipitation of salt in supersaturated brine. The process of precipitation is not instantaneous, as is often the case in liquid-solid precipitation. The process involves the formation of embryos through thermal fluctuations and their subsequent growth, once they achieve stability. With time, more and more precipitates form. This process is called aging. Once the solution achieves an equilibrium composition given by the solvus line for the aging temperature, precipitation stops. The distribution of precipitates affects the hardness and yield strength. The hardness and yield strength are greater when the precipitates are small and finely dispersed in the matrix media than when the precipitates are large and not finely scattered. Therefore, to gain hardness the specimen should be heat treated to produce a fine dispersion of small precipitates. Unfortunately, there is a tendency when thermodynamic equilibrium is achieved for large precipitates to grow and small precipitates to shrink. This lowers the surface to volume ratio of the precipitates, the surface energy, and therefore the energy of the system. At this point, the hardness and the yield point begin to decrease with time of aging. The process of aging is a function of temperature; the higher the temperature, the wider the spacing of the precipitates. They form initially on cooling during solution treatment. In addition, because coarsening is dependent upon the movement of particles, the maximum point is generally reached sooner at a higher temperature than at a lower temperature.

Binary phase diagrams are most commonly used in alloy designing. The simplest binary system is the system that exhibits complete solubility in liquid and solid state (see Fig. 2.1).

The line above which the alloy is in liquid state is called the liquidus line. At temperature just below this line crystals of α solid solution start forming. The line below which solidification goes to completion is called the solidus line. Hence, only "a" and "b" solid solutions exists at any temperature below the solidus line. The intermediate region between liquidus and solidus lines is the two-phase region, where the liquid and solid coexist. The solvus is represented by a line on a phase diagram that separates a solid phase

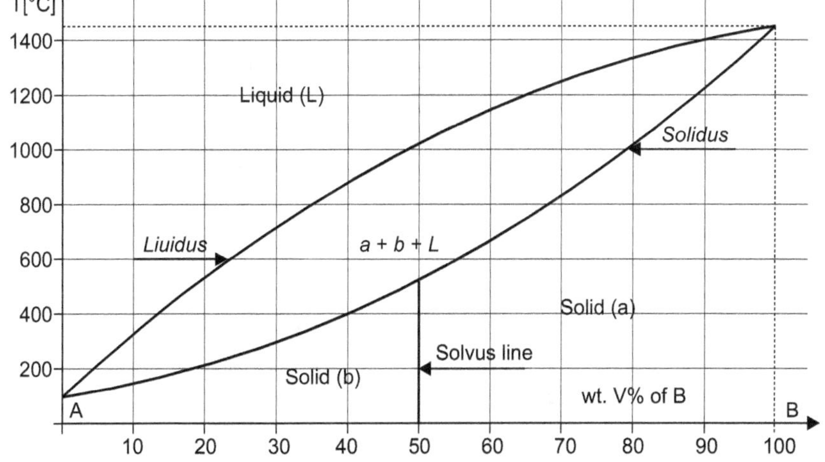

Figure 2.1. Binary phase diagram.

from a solid(a) and solid(b) phase, where solid(a) and solid(b) are different microstructures. It should be noted that the two metals are soluble in each other in the entire range of compositions in both liquid and solid state. This kind of system is known as an 'isomorphous' system. Upon cooling from the liquid state, the temperature of the pure metal (A or B) drops continuously until the melting point at which solidification starts. Solidification happens at a constant temperature. After that, the temperature drops again on completion of solidification. Longitudinal loading and mechanical responses of this type of composite depend on several factors to include the stress-strain behaviors of fiber and matrix phases, the phase volume fractions, and, in addition, the direction in which the stress or load is applied. Furthermore, the properties of a composite having its fibers aligned are highly anisotropic, that is, dependent on the direction in which they are measured. Consider the stress-strain behavior for the situation wherein the stress is applied along the direction of alignment, the longitudinal direction. To begin, assume the fiber to be totally brittle and the matrix phase to be reasonably ductile. Also, the failure stresses in tension for the fiber and matrix are σ_{f*}, and σ_{m*} respectively, and their corresponding fracture strains are ε_{f*} and ε_{m*} furthermore, it is assumed that $\varepsilon_{m*} > \varepsilon_{f*}$ which is normally the case.

A fiber-reinforced composite consisting of these fiber and matrix materials will exhibit the uniaxial stress-strain response; the fiber and matrix behaviors are included to provide perspective. In the initial Stage I, both fibers and matrix deform elastically; normally this portion of the curve is linear. Typically, for a composite of this type, the matrix yields and deforms non-elastically while the fibers continue to stretch elastically, in as much as the tensile strength of the fibers is significantly higher than the yield strength of the matrix. This process constitutes Stage II; this stage is ordinarily very nearly linear, but of diminished slope relative to Stage I. Furthermore, in passing from Stage I to Stage II, the proportion of the applied load that is borne by the fibers increases. The onset of composite failure begins as the fibers start to fracture, which corresponds to a strain of approximately ε_{f*}. Composite failure is not catastrophic for a couple of reasons. First, not all fibers fracture at the same time, since there will always be considerable variations in the failure strength of brittle fiber materials. In addition, even after fiber failure, the matrix is still intact. Thus, these broken fibers, which are shorter than the original ones, are still embedded within the intact matrix, and consequently are capable of sustaining a diminished load as the matrix continues to deform non-elastically. It can also be shown, for longitudinal loading that the ratio of the load carried by the fibers to that carried by the matrix is given by Equation 2.3.

$$\frac{F_f}{F_m} = \frac{E_f V_f}{E_m V_m} \tag{2.3}$$

To illustrate the above conclusions, provide some examples are given below.

Example 2.1—A continuous and aligned glass fiber-reinforced composite consists of 40 vol% of glass fibers having a modulus of elasticity of 69 GPa (psi) and 60 vol% of a polyester resin that, when hardened, displays a modulus of 3.4 GPa (psi).

(a) Compute the modulus of elasticity of this composite in the longitudinal direction.

(b) If the cross-sectional area is 250 mm^2 (0.4 in^2) and a stress of 50 MPa (7250 psi) is applied in this longitudinal direction, compute the magnitude of the load carried by each of the fiber and matrix phases.

(c) Determine the strain that is sustained by each phase when the stress in part (b) is applied.

Solution

(a) The modulus of elasticity of the composite is calculated using Equation 2.1.

$$E_c = 3.4(0.6) + 69(0.4) = 30[\text{GPa}] = [4.3(10^6)\text{psi}]$$

(b) To solve this portion of the problem, first find the ratio of fiber load to matrix load, using Equation 2.3.

In addition, the total force sustained by the composite may be computed from the applied stress and total composite cross-sectional area according to $F_c = 50(250) = 12{,}500$ N. (or: $12{,}500 = F_f + F_m$). Substituting F_f from the above yields: $13.5\ F_m + F_m = 12{,}500$ or: $F_m = 860$ N (200lb) and $F_f =$ 11, 640 N (2700lb).

(c) The stress for both fiber and matrix phases must first be calculated. Then, by using the elastic modulus for each (from part a), the strain values may be determined. For stress calculations, phase cross-sectional areas are necessary:

$$A_m = 0.6(250) = 150\,\text{mm}^2;\ A_f = 100\,\text{mm}^2$$

$$\sigma_m = F_m / A_m = 860/150 = 5.73\text{MPa}\,(833\text{psi})$$

$$\sigma_f = F_f / A_f = 11{,}640/100 = 116.4\text{MPa}(16{,}875\text{psi})$$

Finally,

$$\varepsilon_f = \sigma_f / E_f = 116.4/69000 = 1.69(10^{-3})$$

$$\varepsilon_m = \sigma_m / E_m = 5.73/3400 = 1.69(10^{-3})$$

A continuous and oriented fiber composite may be loaded in the transverse direction; that is, the load is applied at a 90° angle to the direction of fiber alignment. For this situation the stress to which the composite as well as both phases are exposed is the same, or $\sigma_c = \sigma_m = \sigma_f = \sigma$. This is termed an isostress state. In addition, the strain or deformation of the entire composite is,

$$\varepsilon_c = \varepsilon_m V_m + \varepsilon_f V_f .$$

$$\frac{\sigma}{E_c} = \frac{\sigma}{E_m} V_m + \frac{\sigma}{E_f} V_f \quad \text{or:} \quad \frac{1}{E_c} = \frac{1}{E_m} V_m + \frac{1}{E_f} V_f \tag{2.4}$$

This value for E_c is greater than that of the matrix phase but is much smaller than the modulus of elasticity along the fiber direction which indicates the degree of anisotropy of continuous and oriented fiber composites. After selecting the polymer binder, it is necessary to determine the filler, which serves to increase the mechanical properties of nanocomposite. The physical-mechanical characteristics of some nanoparticles are presented in Table 2.1 [2].

However, in many respects, natural materials and designs are still significantly superior to their artificial counterparts. For example, it is known that bird feathers are stronger and substantially lighter than the armor of modern materials such as Kevlar, and in terms of their specific strength and long-term resistance to alternating loads the bird's feather material can compete with modern aluminum magnesium alloys [15,16].

$$\omega_f = A_1 p_1 + B_1 p_1^{C_1}$$
$$\dot{\omega}_c = (1 - \omega_c) 10^{-A_2} \sigma^{B_2} p_2^{C_2} \tag{2.5}$$

where A_i, B_i and C_i (i = 1, 2) are the material parameters.

In order to reduce the number of tests for predicting composite fatigue failure, composite fatigue modeling is required. An interesting article written by Degrieck and Van Paepegem [17] focuses on the existing modeling approaches for the fatigue behavior of fiber-reinforced polymers gives a

Table 2.1. Physical and mechanical characteristics of nanoparticles.

Material	**Al_2O_3**	**BeO**	**Organic-adhesive**	**Nano-adhesive**
Strength limit at tension, [MPa]	300	-	89.6	101
Strength limit at bending, [MPa]	400	-	-	145
Modulus of elasticity, [GPa]	370	317	4.83	4.60
Density, [g/cm3]	3.96	2.30	1.15	1.90
SAP Components	6...8%	σ_b, [MPa] = 300	$\sigma_{0.2}$[MPa] = 220	Aver. = 7%

comprehensive survey of the most important modeling strategies for fatigue behavior. A more recent paper written by Sevenois and Van Paepagem [18] gives an overview of the existing techniques for fatigue damage modeling of FRPs with woven, braided and other 3D fiber architectures. The aim of the present chapter is not to give an in-depth discussion of the fatigue models; thus, the interested reader will be asked to refer to references [17, 18]. In the first reference, the authors justify the classification, currently made by Sendeckyj et al. [19], concerning the large number of existing fatigue models for composite laminates. This classification consists of three major categories: fatigue life models (empirical/semi-empirical models), which do not take into account the actual degradation mechanisms, but use S-N curves or Goodman-type diagrams introducing a fatigue failure criterion; phenomenological models for residual stiffness/strength; and, finally, progressive damage models (or mechanistic models), which use one or more damage variables related to observable damage mechanisms (such as transverse matrix cracks and delamination). Note that this classification has been recently slightly modified for fatigue damage modeling techniques for FRP (Fiber Reinforced Polymers) with woven, braided or other 3D fiber architectures [20], but the classification is reported in [19]. Obviously, it is difficult to get a general approach of the fatigue behavior including polymer matrix, metal matrix, ceramic matrix composites, elasto-metric composites, glare, short fiber reinforced composites and nanocomposites. The failure in high performance composite material structures is of quite a different nature from that of metallic components. Failure, due to fatigue, is to be feared much less in the first case than in the second.

Metallic fatigue has been intensively studied for more than 150 years. Nowadays the fatigue life of composites is approached using some well-established concepts:

- Low cycle fatigue from L. Coffin and Manson
- Mega cycle fatigue from A. Whoeler
- Giga cycle fatigue from C. Bathias
- Crack propagation from P. Paris

The S – N curves of composite materials usually have between 10^3 and 10^6 cycles before failure. The delamination, which in the first approximation implies flaws in the plane, can be treated by damage mechanics using the modified Paris law. In some cases where a quasi-isotropic lay-up exists, it is always possible to use damage mechanics to predict the failure of notched plates. Other criteria such as Nuismer and Whitney approach [21] or damage mechanics models [22] are recommended. In composites materials, the fatigue damage is multidirectional and the damage zone, much larger than the inelastic zone, is related to the complex morphology of the fracture [23].

According to the difference of fatigue mechanics, the fatigue laws are not the same. It is well-known fact that the endurance curve of metallic alloys is quasi-hyperbolic in shape, with a pronounced concavity as soon as the maximum stress of the cycle exceeds the elastic limit of the material. The concavity of the endurance curve is attributed to the physical of composites. Below 10^5 cycles, when inelasticity is generalized, the lifetime N_f is given by the Manson-Coffin relation: in which ε and ε_p are the elastic and inelastic strains. Considering that ε_p is negligible in high performance composites, the lifetime will be expressed by a relation of the form: $\varepsilon = 2\beta[2N_f]^c$, where β and c are composite material parameters.

2.5 Cyclic loading types

Constant minimum and maximum stress levels are referred to as constant amplitude loading. This is a much simpler case and will be discussed first. Otherwise, the loading is known as variable amplitude or non-constant amplitude and requires special treatment. Fully-reversed loading occurs when an equal and opposite sign load is applied. This is a case of $\sigma_m = 0$ (zero medium stress) and $R = -1$. A zero-based loading case occurs when a load is applied and removed. This is a case of $\sigma_m = \sigma_{max}/2$ and $R = 0$. When doing a test on the stress range versus fatigue life and plotting the results on two logarithmic axes, the results tend to be linear. Basquin formulated this mathematically in a power law. The curve may consist of several linear pieces. Two parameters for each of the curves in Basquin's law are needed as input: Starting point S_e and slope b. The properties are usually found for zero mean stress and a uniaxial tension stress state on polished specimens. Basquin's law [24,25] is usually presented as,

$$N_f = N_e \left(\frac{S}{S_e} \right)^{1/b} \qquad (2.6)$$

Mean stress correction

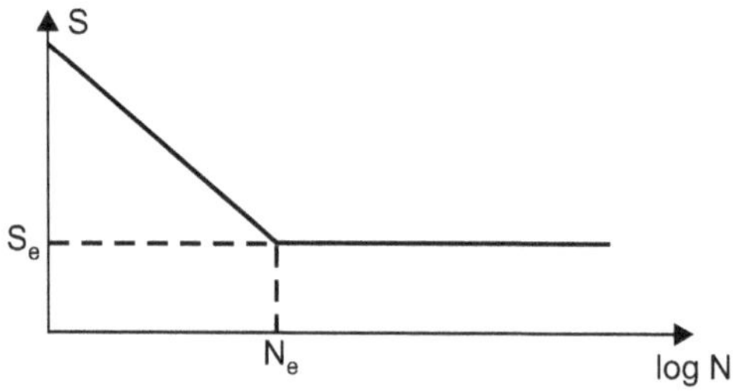

Figure 2.2. Basquin's law.

For fatigue loadings that have small mean stress compared to the alternating stress, the theories show little difference. Goodman law [20] is presented as:

$$\left(\frac{\sigma_a}{\sigma_{a'}}\right)+\left(\frac{\sigma_m}{\sigma_u}\right)=1$$

$$\sigma_{a'}=\frac{\sigma_a}{1-\left(\frac{\sigma_m}{\sigma_u}\right)} \tag{2.6a}$$

$\sigma_{a'}$ = Corrected stress amplitude; σ_a = Initial stress amplitude; σ_m = Mean stress

σ_u = Ultimate Tensile Strength, UTS

Similarly, Gerber law **[26]** is presented as:

$$\sigma_{a'}=\frac{\sigma_a}{1-\left(\frac{\sigma_m}{\sigma_u}\right)^2} \tag{2.7}$$

Finally, we have the same for the Soderberg law [27]:

$$\sigma_{a'}=\frac{\sigma_a}{1-\left(\frac{\sigma_m}{\sigma_y}\right)} \tag{2.8}$$

Strain life analysis, theory

- This analysis is better suited than a Stress Life analysis to cope with higher stress ranges on the model because it contains an additional term compared to the Stress-Life analysis and it also offers correction models for local inelasticity.
- The strain-life method is suitable in the lower cycle fatigue range involving less than about 1.000 to 10.000 cycles fatigue life and high local stress (So called low cycle fatigue, LCF).

Basquin-Coffin-Mansons law

The B-C-M law calculates the number of cycles to failure as:

$$N_f = f\,(\sigma_f, b, \varepsilon_f, c) \tag{2.9}$$

σ_f = Fatigue strength coefficient: b = Fatigue strength exponent: ε_f = Fatigue ductility coefficient c = Fatigue ductility exponent: E = Young's modulus

In this case Equation 2.9 is presented as:

$$\frac{\Delta\varepsilon}{2} = \frac{\sigma_f}{E}\left(2N_f\right)^b + \varepsilon_f\left(2N_f\right)^c \tag{2.10}$$

The Morrow model accounts for mean stresses by moving the elastic part of the material curve up and down according to the mean stress of each cycle.

$$\varepsilon_a = \frac{(\sigma_f - \sigma_m)}{E}\left(2N_f\right)^b + \varepsilon_f\left(2N_f\right)^c \tag{2.11}$$

Smith-Topper-Watson accounts for mean stresses by using a damage parameter gathered from the maximum strain at each cycle.

$$\varepsilon_a = \frac{1}{\sigma_{max}}\left[\frac{\sigma_f}{E}\left(2N_f\right)^{2b} + \varepsilon_f\left(2N_f\right)^{b+c}\right] \tag{2.12}$$

Smith-Topper-Watson law should be used for loads involving tensile stresses, whereas Morrow law is best suited for those involving compressive stresses. Since fatigue of metallic materials is a well-known phenomenon, first attempts to account for fatigue in composites consisted in adapting the already existing methods for metallic materials to composites.

2.6 Creep-fatigue constitutive model with cumulative damage law

The constitutive equations used in this study are based on the additive decomposition of the strain rates into their elastic ε_e, inelastic ε_{in}, and thermal/chemical $\varepsilon_{th/ch}$ expansion parts. The total strain rate is given as

$$\dot{\varepsilon} = \dot{\varepsilon}_e + \dot{\varepsilon}_{in} + \dot{\varepsilon}_{th/ch} \tag{2.13}$$

For the elastic part, the stress tensor is related to the elastic strain and the expression is given as

$\bar{\sigma}_e = E\varepsilon_e$—Instantaneous stress for the elastic partand the stress-strain behavior is studied because there was no damage and it is assumed that the fatigue strain is: $\varepsilon_e = a_0 \sin(pt)$. Damage is considered when a material or a structure generates inelastic deformations. An integral type viscoelastic equation with instantaneous stress is used to describe the evolution of the creep-fatigue nonlinear inelastic process which has the following form:

$$E[T(t)][a_0 \sin(pt)] = \sigma(t) + \int_0^t e^{\frac{E_a}{RT}} \frac{1}{(1-\omega)^r} F(t,\tau)\sigma^n(\tau)d\tau + \sigma_{st} =$$

$$= \sigma(t) + \int_0^t A_1 e^{\frac{E_a}{RT}} \frac{1}{(1-\omega)^r}]K_1(t,\tau)\sigma^n(\tau)d\tau +$$

$$+\int_0^t A_2 e^{\frac{E_a}{RT}} \frac{1}{(1-\omega)^r}]f_2[\sigma(\tau)]K_2(t,\tau)\sigma^n(\tau)d\tau +$$

$$+\int_0^t \frac{E_0}{E_1} A_3 e^{\frac{E_a}{RT}} \frac{1}{(1-\omega)^r} f_3[d(\tau)]K_3(t,\tau)\sigma(\tau)d\tau + \sigma_{st}$$

$$\frac{d\omega}{dt} = A\frac{\sigma^q}{(1-\omega)^r}; \quad \omega(0) = \mathbf{0}; \quad \sigma(\mathbf{0}) = \mathbf{0}$$

$$K_1(t,\tau) = \varphi_1(t)f_1(\tau) = \sum_{i=1}^{N}[\sin p(t-\tau)]\exp(-\alpha_i t)\exp(\alpha_i \tau) \tag{2.14}$$

$$K_2(t,\tau) = \sum_{i=1}^{N}[\sin p(t-\tau)]\exp(-\beta_{i2}t)\exp(\beta_{i2}\tau)$$

$$K_3(\theta,\tau) = \sum_{i=1}^{N}[\sin p(t-\tau)]\exp(-\beta_{i3}t)\exp(\beta_{i3}\tau)$$

$$E = \left(a - b\left(\tanh\left(c\left(T - T_{gm}\right)\right)\right)\right)(1-\varphi) + \left(a - b\left(\tanh\left(c1\left(T - T_{gf}\right)\right)\right)\right)\varphi / k$$

$$f_2(\sigma) = \sigma^s; \quad s = 1,2,3\ldots,M;$$

$k = E_0/E_1$; $A_1 = \eta_0/E_0$; $\text{ç} = \mathbf{exp(-b_1(T - T_*)}$ – Reynolds formula

$$A_2 = A_1(\exp(-b_1T))((1+2.5\varphi+14.7(\varphi^2))d_{eff1}$$

$$d = a_1T; \quad A_3 = [f_3(T)][d_{eff1}(T)]; \quad d_{eff} = (\varphi d)/(1 + 0.333(1 - \varphi)d)$$

$$d_{eff1} = (1+fi*((10*t+1)\wedge 0.333-1))\wedge 3 - 1$$

In integro-differential Equation (2.14)—a; a_1; b; b_1; c; c_1; A_1; A; E_0; E_1; k; n; s; q; r; β_{i1}; β_{i2}; β_{i3}; η_0 are constants. T – temperature [°F]; t – time [sec]; T_{gm} and T_{gf} – matrix and filler glass transitional temperatures respectfully; $\varphi = V_f/(V_m + V_f)$ – volumetric concentration of filler; E_a – activation energy; R – universal gas constant; d – macro scale dimension of nanoparticle cluster (or dispersed filler material);

d_{eff} – macro scale dimension of cluster (Maxwell Garnet equation (MG) [28]);

d_{effl} – macro scale dimension of cluster (Landau and Lifshitz (1984) [29]); σ is creep-fatigue stress; p – arbitrary oscillation frequency. The first integral on the right-hand site of the Equation (2.14) represents the accumulated viscoelastic stress due to effect of high external thermal load. The second integral represents the accumulation of creep-fatigue stress affected by the combination of internal stress and high temperature. Its evolution, characterized by a flow rule, is of primary importance in describing viscoelastic behavior of composites and nanocomposites. The third integral represents the effect of growing clusters and nanoparticles and free chemical energy in the creep-fatigue process. The influence of the continuum damage variable ω and the effect of high thermal load are also considered in this integral. The evolution of creep-fatigue process in this case is presented by including the laws of *the kinetics of a phase transformation in the material in the integrand.* It should be noted that the mechanical parameters of the composite material also depend on its chemical composition and the technological method of its production, and hence on the physical kinetics of the chemical reaction itself (at least in a simplified final form). This in turns determines ultimately the temperature-time dependence (different from the temperature-time dependence of the external thermal load) affecting the creep-fatigue process of composites and nanocomposites. This is done by introducing the function f3 in Equation (2.14) (analogy with the well-known HP principle [11,12]. Functions are introduced for the stress dependence on the size of nanoparticles and so-called the degree of crystallization. The best description of the kinetics of phase transformation is given by the so-called Kolmogorov-Johnson-Mehl-Avrami (KJMA) theory [13–17].

The knowledge of the kinetics of crystallization of nanomaterials is a key point in constructing the creep-fatigue model that depends on the microstructural scale. However, from the basic research point of view, it helps to validate the proposed models for phase transformations. *A* monotonically increasing (decreasing) function f3 (T) is a solution of the differential equation as follows:

$$\frac{d(f3)}{dT} = f(f3, T) \tag{2.15}$$

In general, the analysis of this differential equation is based on two main assumptions:

(1) The transformation rate at temperature T during a chemical reaction, df3/dT, can be expressed as a product of two separable functions, one depending solely on the temperature, T, and the other solely on the fraction transformed f3.

(2) The temperature dependent function, called the rate constant, follows an Arrhenius type dependency

The Equation (2.15) has the form:

$$\frac{d(f3)}{dT} = Ae^{-\frac{E}{RT}} f(f3) \tag{2.16}$$

Equation (2.16) is the crystallization kinetics equation. In it, T denotes the temperature at the current point in space at the current time. The distribution of temperatures in a continuum medium is found from the solution of the energy conservation equation (see earlier author's work [29]):

$$\rho \frac{\partial(cT)}{\partial t} = -\mathrm{div}\vec{q} + q \tag{2.17}$$

where, ρ *is the density*; c is the specific heat of the material, and q is the specific heat release power. The heat flux vector q is related to the temperature gradient by the Fourier law,

$$q = -k_{ij} \frac{\partial T}{\partial x_j} \tag{2.18}$$

On the right-hand side of the Equation (2.18) k_{ij} is the thermal conductivity tensor k_{ij}. Due to the chemical reaction during the curing process, heat is produced and it is proportional to the rate of *crystallization of nanomaterials*. For the nanocomposites analysis used in this work, the *crystallization* kinetics equation was taken as the n-th order Prout-Tompkins equation [30] with autocatalysis effect:

$$\frac{dC}{dt} = Ae^{-\frac{E}{RT}} (1 - C)^q C^p \tag{2.19}$$

A, q, p – are constant parameters in Equation (2.19)

The heat Equation (2.19) at hand is also separable. This means that the mechanical deformation of the structure does not affect the heat transfer processes occurring in it. For further detailed discussions on this matter see reference [31]. Some types of functions f3 are presented below in the numerical analysis. It is assumed here that the material parameters are based on simple experimental data.

Finally, the dimensionless form [12] of Equation (2.14) is as follows:

$$m = 0.0405\theta - 0.01126\theta^2 + 0.001462\theta^3 - 0.00006868\theta^4$$

$$m1 = 0.0405 - 0.02252\theta + 0.004386\theta^2 - 0.0002747\theta^3$$

$$E[\theta][a_0 \sin(pm)] = \sigma_0(\theta) + \int_0^\theta A_1 e^{\frac{\theta}{1+\beta 1\theta}} \frac{1}{(1-\omega)^r} K_1(\theta,\tau)\sigma^n(\tau)m\mathbf{1}d\tau +$$

$$+\int_0^\theta A_2 e^{\frac{\theta}{1+\beta\theta}} \frac{1}{(1-\omega)^r} f_2[\sigma(\tau)]K_2(\theta,\tau)\sigma^n(\tau)m\mathbf{1}d\tau +$$

$$+\int_0^\theta \frac{E_0}{E_1} A_3 e^{\frac{\theta}{1+\beta\theta}} \frac{1}{(1-\omega)^r} f_3[d(\tau)]K_3(\theta,\tau)\sigma(\tau)m\mathbf{21}d\tau + \sigma_{st}$$

$$\frac{d\omega}{d\theta} = A\frac{m\sigma^q}{(1-\omega)^r} + A_0\frac{m\mathbf{21}\sigma^q}{(1-\omega)^{r1}}; \;\; \omega(0) = \mathbf{0}; \;\; \sigma(\mathbf{0}) = \mathbf{0}$$

$$K_1(\theta,\tau) = \varphi_1(\theta)f_1(\tau) = \sum_{i=1}^{N}[\sin[p(\theta-\tau)]\exp(-\alpha_i m)\exp(\alpha_i m)$$

$$K_2(\theta,\tau) = \sum_{i=1}^{N}[\sin p(\theta-\tau)]\exp(-\beta_{i2}m)\exp(\beta_{i2}m) \tag{2.20}$$

$$K_3(\theta,\tau) = \sum_{i=1}^{N}[\sin[p(\theta-\tau)]\exp(-\beta_{i3}m\mathbf{2})\exp(\beta_{i3}m\mathbf{2})$$

$$E = \left(a - b\left(\tanh\left(c\left(\theta - \theta_{gm}\right)\right)\right)\right)(\mathbf{1}-\varphi) + \left(a - b\left(\tanh\left(c\mathbf{1}\left(\theta - \theta_{gf}\right)\right)\right)\right)\varphi / k$$

$$f_2(\sigma) = \sigma^s; \;\; s = \mathbf{1},\mathbf{2},\mathbf{3}..., N;$$

$$k = E_0/E_1; \; A_1 = \eta_0/E_0; \; \text{ç} \; = \; \mathbf{exp(-b_1}\,(\beta\theta)) \; \text{– Reynolds formula}$$

$$A_2 = A_1(\exp(-b_1\beta\theta))((1+2.5\varphi+14.7(\varphi^2))d_{eff1}$$

$$d = a_1\theta; \;\; A_3 = [f_3(\theta)][d_{eff1}(\theta)]; \;\; d_{eff} = (\varphi d)/(1 + 0.333(1 - \varphi)d)$$

deffl= (1+φ*((10*t+1)^0.333-1))^3 - 1

$$m2 = 0.0868\theta - 0.0386(\theta^2) + 0.0080(\theta^3) - 0.000756(\theta^4) + 2.6(10^{-5})(\theta^5)$$

$$m21 = 0.0868 - 0.0772\theta + 0.024(\theta^2) - 0.003(\theta^3) + 1.3(10^{-4})(\theta^4)$$

$$\beta = \frac{RT_*}{E_a}; \;\; T = \beta T_*\theta + T_*; \;\; T_* - \text{ Base temperature } [{}^0K]$$

Before proceeding to a detailed analysis of the fatigue behavior and the life span of composites and nanocomposites based on solutions of integral-differential equations of system (2.20), it is natural to make sure that unique solutions to Equation (2.20) *exist*. The existence and uniqueness theorems for the solutions of integral-differential equations are the foundation on which all kinds of calculations of nonlinear equations are based. The subject of nonlinear integral equation is considered as an important branch of mathematics, therefore the exact proof of the corresponding theorem is not provided here. However, two things have to be underlined here with respect to the uniqueness of solutions of Equation (2.20): (1) All functions that are included in Equation 2.20 are analytical, continuous and integrable. (2) Any differential and integral (linear and nonlinear) equations must have initial conditions, therefore $\omega(0) = 0$ and $\sigma(0) = 0$ are added. It is assumed also that any composite material consists in general of two phases: primary that forms the matrix and secondary–referred to as the embedded phase (reinforcement, fibers, particles, and more). The current constitutive model is formulated in a small strain framework, thus we can consider the additive decomposition of the total strain tensor. This treatment can simplify the model and facilitate the future numerical calculations. The Young's modulus E_0 can be identified from the slope of the linear region of the stress-strain curves during the uniaxial static load tests. It is assumed here that the modulus of elasticity function, E, is explicitly a function of temperature only. However, as noted above, the function of the external temperature - time is considered to be given, then we can say that the function of the modulus of elasticity in Equation (2.20), E, is also a function of time. For example, the functions m and m1 are the inverse functions of the temperature-time dependence in the event, taken from [29]. Within the framework of the current model, the value of F_y can be determined by using the uniaxial loading-dwell tests, which just equals the limiting value of the total stress after a long holding time. Creep strength decreases and both creep strain and creep rate increase with increasing temperature. The combination of creep and fatigue damage mainly at time varying and elevated temperatures is typically more damaging than the additional damage caused by each individual loading condition. Thermo-mechanical creep-fatigue tests are very complicated and expensive, therefore a phenomenological model is preferred to predict fatigue life under high temperature loading conditions. Empirical relations have been developed to represent the temperature and strain rate dependencies of composite properties at creep-fatigue and failure stages. At temperatures below T_g or at high strain rates, fiber pullout and matrix brittle fracture are dominant failure mechanisms, while at temperatures above T_g or low strain, failure occurs via matrix crazing and crack propagation. Creep-fatigue resistance decreases with increasing temperature. The time-temperature-stress superposition procedure has been used for extending short-term creep data for long-term creep predictions in most of the studies. Fatigue strengths at high temperatures must be quoted, not as fatigue limits, but as stress limits that can be withstood for a certain number of stress cycles without failure [34].

The constitutive equations of the current creep-fatigue model are as follows. The total strain tensor is supposed to be composed of the elastic, thermal and inelastic strains. We assume that isotropic heat expansion and physical inelasticity contribute only to the deviatoric strain. The elastic strain and stress tensor satisfy the integral type of nonlinear creep law, where the instantaneous bulk modulus of elasticity is denoted as E (T) and is a function of current temperature only. Following the author's reference framework [35], the total inelastic strain rate is decomposed into the viscoelastic strain rate and the creep strain rate. For the viscoelastic strain rate, the exponential viscosity function is adopted from reference [36], where the time variant viscous equivalent stress σ depends on the continuum damage tensor, the isotropic hardening variables and the initial instantaneous strain as a periodic function with an arbitrary frequency. The Kachanov-Rabotnov [9] type kinematic equation with two components is adopted, where the first term is chosen as a damage parameter affecting the 'regular' composites and the second one represents the damaged portion for 'dispersed' types of composites and nanocomposites. Creep and stress relaxation properties are vital to material and engineering design because they are related to the composite molecular structure and associated chemistries in the formulation. The chemical aspects of composite formulations affecting mechanical performance and design include:

- The demand for a fast turnaround in the design of new composite and nanocomposite materials necessitates quick completion of associated laboratory tests, which includes mechanical characterization. Consequently, it would not be practical to perform multiple creep experiments on composite 'A' using many different temperatures and load conditions.
- Under the conditions of creep-fatigue, in general, three microstructural mechanisms cause the accumulation of damage and inelastic strain: the *nucleation and growth* of internal and grain boundary cavitations, the slip and climb of dislocations and the formation of persistent slip bands during cyclic loading. These microstructural mechanisms are strongly correlated to creep, and fatigue damage, ω_c, and ω_f respectively. The purpose of the damage evolution equations are to model the softening phenomena observed in mechanical tests and track the evolution of microstructural defects.

Creep damage is represented by homogenous nucleation, growth, and coalescences of cavities near the end of fatigue life. A large number of micro-voids nucleate on grain boundaries. Creep damage produces the tertiary creep regime. During monotonic tension, ductile softening is observed after the ultimate tensile strength has been exceeded. This inelastic-ductile damage is represented by the rapid (when compare to creep damage) nucleation, growth, and coalescences of micro voids. Inelastic damage produces necking in specimens and is often characterized by some critical strain or stress. During strain-controlled fatigue, softening occurs in the stress amplitude in cycles preceding cyclic-stress saturation. This fatigue damage occurs due to crack nucleation, growth, and fast fracture where cycling creates striations and extension in the crack length.

2.7 Proposed constitutive model of the creep-fatigue process

Under creep-fatigue, both creep and cyclic viscoelasticity contribute to the fatigue evolution of composite material. This necessitates the development of a unified mechanical model for the creep-fatigue process:

- Primary, secondary, and tertiary creep
- Monotonic and cyclic viscoelasticity
- Hardening, softening, and saturation
- damage and failure prediction
- mechanical and microstructural degradation

In this "unified" mechanical model all inelastic strains are derived from a single viscous function. Creep and viscoelasticity are independent of each other; rather the state variables of the viscous function evolve such that both phenomena can be modeled. The development of the unified mechanical model involves following steps. First the appropriate state variable must be selected. Next, a viscous function must be found which models the secondary creep behavior and incorporates steady-state values of the state variables. Then, the primary creep, monotonic tensile and cyclic behavior can be modeled by converting the state variables into evolving functions. Then, coupled CDM-based creep and fatigue equations must be developed to model tertiary and post cyclic stress saturation behavior towards the prediction of failure. Finally, the influence of the damage function on the mechanical properties and microstructure of the material must be investigated.

Dispersed Composites and Nanocomposites

The development of a unified mechanical model for creep involves many steps. First, an appropriate viscous function which relates the minimum creep strain rate to stress must be determined for the subject material. This function must incorporate the ability to deal with temperature-dependence. Next, a damage evolution equation must be generated and coupled with the viscous function. This coupling must be done in such a way that the viscous function reverts back to its original form when damage is zero. The damage evolution equation must be formulated to mitigate stress-sensitivity and replicate the evolution of microstructural defects. Afterwards, a "special" equivalent stress must be found to incorporate the issue of multiaxiality. This "special" equivalent stress is incorporated into the viscous and damage evolution equations. The proposed constitutive model is multistage where primary, secondary, and tertiary creep regimes are modeled. This is done by separation into primary and secondary viscous functions as follows:

$$\dot{\varepsilon}_{cr} = \dot{\varepsilon}_{pr} + \dot{\varepsilon}_{sc}(\omega) \tag{2.21}$$

where, the tertiary regime arises from the damage variable, ω. The secondary viscous function must be found. The form of the secondary viscous function is given by the relationship between the minimum creep strain rate and stress as follows:

$$\dot{\varepsilon}_{min} = f(\sigma)g(T) \tag{2.22}$$

The viscous function must be modified to incorporate temperature-dependence. This is done by making the material constants functions of stress or creep coefficient g(T). In case of the creep coefficient, g(T), a common approach is to use an Arrhenius type relation as follows

$$g(T) = A_0 \exp\left(-\frac{E_a}{RT}\right) \tag{2.23}$$

where A_0 (MPa^{-1} hr^{-1}) is the pre-exponential factor, E_a (J mol) is the apparent activation energy, R is the universal gas constant 8.314 [$Jmol^{-1}$ K^{-1}], and T(^{0}K) is temperature [33]. The main goal here is to determine the stress-strain relationship ($\sigma - \varepsilon$) and the damage function X in terms of the given instantaneous strain and temperature – time relationships. The main parameters that significantly affect the behavior of composite and nanocomposite materials at high temperatures are:

- Load amplitude: Initial (static) stress level: Mean Stress: Temperature: Frequency: Hold time: Waveform and R values: Material Processing: Influence of volumetric percentage of filler material on service life and residual load bearing capacity of a structure: Rate of transitional temperature change, T_g (or θ_g), are a function of temperature and viscosity: Ratio $k = E_f/E_m$ affects the service life and residual load bearing capacity of a composite structure and the Nonlinear exponent n (or s) of the stress power law affects the service life and residual load bearing capacity of a structure.

At the same time, it is necessary to emphasize once again that the ultimate goal of the proposed model is the construction of the stress-number of cycles before fatigue failure (lifetime of the composite structure). Since the problem of fatigue creep of composites in the most general case is extremely complex and multifaceted from the point of view of the diversity of the micro- and macro-structure of such materials, to achieve this main goal, three basic conditions are needed when building any phenomenological fatigue-creep model.

(1) Possible simplification of the task at the macroscopic level (from the physical and physicochemical aspects of the issue;
(2) The maximum possible reduction in the number of parameters used in this model, obtained on the basis of relatively simple experiments (creep and fatigue);
(3) The development of approximate methods for numerical analysis of the fatigue-creep model of composites with the possibility of further improvement and verification of the results obtained by comparison with the available experimental data.

The main assumptions are as follows:

1. The instantaneous strain and temperature – time relationships (in dimensionless form) are given. The temperature – time relationship, for example, might be provided as a solution of the simple kinetic equation presented below A↔ P [36]:

Example 2.1—Data: $k^+ = 12$; $k^- = 1$; P – product (nanoparticles'); $y = \boldsymbol{\theta}$ – dimensionless temperature

$$\begin{aligned}
&t(\mathbf{0}) = 0;\ t(f) = \mathbf{0.12}\\
&d(P)/d(t) = \mathbf{4}*(\mathbf{12}*A - \mathbf{1}*P)\\
&P(\mathbf{0}) = \mathbf{0}\\
&d(A)/d(t) = \mathbf{4}*(-\mathbf{12}*A + \mathbf{1}*P)\\
&A(\mathbf{0}) = \mathbf{1}\\
&d(y)/d(t) = \mathbf{20}*P\wedge\mathbf{1}*\exp(y/(\mathbf{1}+.\mathbf{1}*y)) - \mathbf{0.233}*y\wedge\mathbf{4}\\
&y(\mathbf{0}) = \mathbf{0}\\
&\varepsilon_{inst} = a_0\sin(pm)
\end{aligned} \tag{2.24}$$

Calculated values of DEQ variables

	Variable	Initial value	Minimal value	Maximal value	Final value
1	A	1.	0.0787229	1.	0.0787229
2	P	0	0	0.9212771	0.9212771
3	t	0	0	0.12	0.12
4	y	0	0	11.12286	11.12286

Differential equations

1 d(P)/d(t) = 4*(12*A-1*P)

2 d(A)/d(t) = 4*(-12*A+1*P)

3 d(y)/d(t) = 20*(1-P)^0*P^1*exp(y/(1+.1*y)) – 0.233*y^4

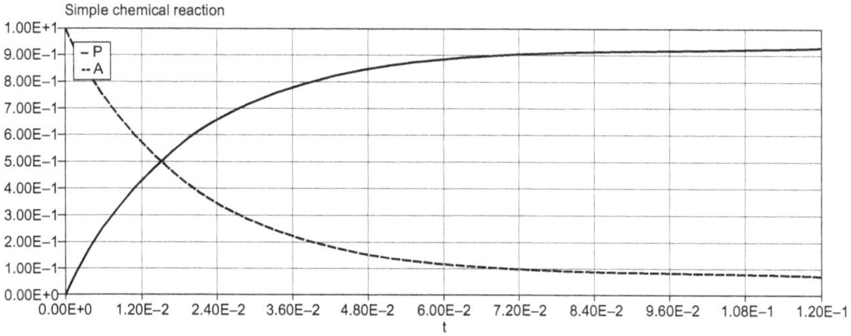

Figure 2.3. Simple kinetic chemical reaction.

Model (reversed function is: t = a1*y + a2*y^2 + a3*y^3 + a4*y^4 + a5*y^

Variable	Value
a1	0.0868221
a2	–0.0385558
a3	0.0080312
a4	–0.0007564
a5	2.609E-05

$$\tau = 0.0868(\theta) - 0.0386(\theta^2) + \mathbf{0.0080(\theta^3) - 0.000756(\theta^4) + 2.6(10^{-5})\ (\theta^5)} \quad (2.25)$$

y = $\boldsymbol{\theta}$; t = $\boldsymbol{\tau}$

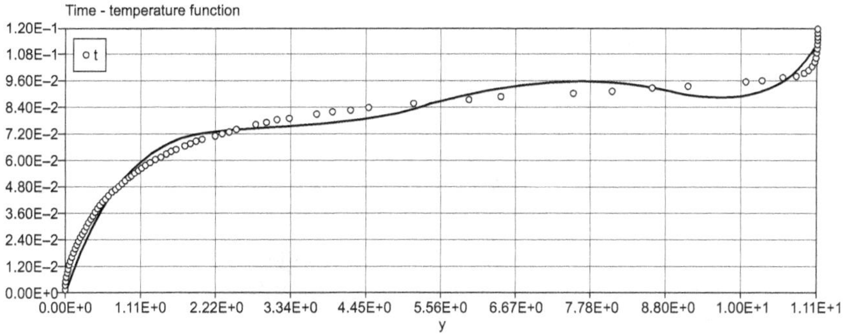

Figure 2.4. Time – temperature function.

$$\varepsilon_{inst} = a_0 \sin(pm) \quad \text{or:} \quad \varepsilon_{inst} = a_0 \sin(pt) \rightarrow a_0 \sin\left(\frac{p}{k}\tau k\right) \rightarrow a_0 \sin(pm)$$

$$m = 0.01\theta$$

$$m1 = 0.01 \tag{2.26}$$

$$m2 = 0.0868\theta - 0.0386(\theta^2) + 0.0080(\theta^3) - 0.000756(\theta^4) + 2.6(10^{-5})(\theta^5)$$

$$m21 = 0.0868 - 0.0772\theta + 0.024(\theta^2) - 0.003(\theta^3) + 1.3(10^{-4})(\theta^4)$$

$$\varepsilon_{inst} = a_0[\sin(pm)]; a_0 = a / a_{st}; \varepsilon_{inst} = a_0[\sin(pt)]; a_{st} = \sigma_0 / E_0$$

2.8 Analytical formulas for S – N fatigue curves

Most of the phenomenological theories of creep are built on the test results of material samples for uniaxial creep at several levels of constant stress and uniaxial cyclic fatigue loads at several levels of cyclic amplitudes and frequencies. The simplest way to check the validity of the creep-fatigue phenomenological equations is the comparison with the corresponding experimental testing data of the parameters. As one of such loading modes, the regime of pure cyclic loading/unloading ($R = -1$ and $a_{st.} = 0$) is considered, and the integral equation of fatigue-creep is supplemented by the differential equation of the cumulative damage accumulation in the composite or nanocomposite. To solve such problems of cyclic creep and fatigue of structures made of nonlinear viscoelastic materials, Rabotnov's hereditary theory of creep is often used [9]. The governing equations of the theory are obtained proceeding from the condition of similarity of isochronous creep diagrams and the cyclic unloading mode. In [4], the diagrams of creep and fatigue (S – N fatigue curves of the composite) were extended by introducing the experimentally obtained instantaneous deformation function into the constitutive Equation (2.14). This chapter solves the problem of calculating the deformations of nonlinear viscoelastic composites under uniaxial loading based on the model built in [4]. This chapter presents two different methods for life prediction of a composite at high temperature; namely continuum damage mechanics (CDM) and analytical composition of S – N creep-fatigue curves. Let's start by plotting fatigue curves (S – N curves), provided that the frequency p = 100 is a constant value and the stress amplitude (a0 = 0.15, 0.25, 0.35 ...., 0.95). Thus, solving Equation (2.14) for each value of a_0, the maximum value of the dimensionless temperature is found, at which the condition $\omega(t) = 1$ is satisfied. It is easy to see that $N_f = pm/2\pi$. Thus, a point on the composite fatigue curve is obtained. Repeating this calculation procedure for each value of a_0, we obtain the approximate values of the composite life span. The calculation examples below are self-explanatory. Using POLYMATH software [37] we have:

Example 2.1a—p = 100; $a_0 = 0.15$

Calculated values of DEQ variables

	Variable	Initial value	Minimal value	Maximal value	Final value
1	A	2.	2.	2.	2.
2	A0	2.	2.	2.	2.
3	A1	1.	1.	1.	1.
4	A2	0	0	0.0035422	0.0003444
5	A3	0	0	0.2007966	0.2007966
6	c1	3.	3.	3.	3.
7	c2	5.	5.	5.	5.
8	d	0	0	10.5388	10.5388
9	deff	0	0	0.1215843	0.1215843
10	deff1	0	0	0.2007966	0.2007966
11	E	1.45	0.3625	1.45	0.3625
12	f3	1.	1.	1.	1.
13	f31	0	−1.672248	0.999998	−1.672248
14	f32	1.	0.6560277	1.	0.6560277

$$X = \omega = 0.12\theta - 0.118\theta^2 + 0.0373\,\theta^3 - 0.00467\theta^4 + 0.0002\theta^5 \quad (2.27)$$

Thus, it is clear that the criterion for fatigue of a composite (and also at the same time a condition for failure) is the equality accepted here (as well as in many other cases for life prediction at high temperature; namely continuum

15	f33	0	0	0.999929	0.8344902
16	f34	0	0	1.735413	1.735413
17	f35	1.	0	1.	0
18	f36	0	0	1.	1.
19	fi	0.05	0.05	0.05	0.05
20	k	10.	10.	10.	10.
21	m	0	0	0.105388	**0.105388**
22	m1	0.01	0.01	0.01	0.01
23	m2	0	0	0.0825478	0.0459426
24	m21	0.0868	0.0007311	0.0868	0.0309236
25	n	1.	1.	1.	1.
26	p	100.	100.	100.	100.
27	s	1.	1.	1.	1.
28	t	0	0	11.	**11.**
29	t1	4.	4.	4.	4.
30	t2	6.	6.	6.	6.
31	X	0	0	0.9986549	**0.9986549**
32	X1	0	0	0.8284823	0.8284823
33	X2	0	0	0.1701726	0.1701726
34	Y1	0	–0.1240769	0.2174998	–0.0488006
35	z	0	–27.97411	0.2199015	–27.97411
36	z1	0	–0.0177155	28.30993	28.30993
37	z11	0	–0.6662722	1.132E+10	1.132E+10
38	z12	0	–0.2079645	5.564E+09	5.564E+09
39	z2	0	–0.113014	0	–0.113014
40	z21	0	–8.802E+07	0.0001213	–8.802E+07
41	z22	0	–4.326E+07	2.689E-05	–4.326E+07
42	z3	0	–6.183E-05	0.0508749	0.024107
43	z31	0	–3.653724	0.0005033	–3.653724
44	z32	0	–0.4093573	0.0179208	–0.4093573

Differential equations

```
1   d(z11)/d(t) = (1/(1-X)^6)*(exp(t/(1+0.067*t)))*A1*((exp(0.1*m)))*m
    1*(sin (p*m))*z^n
2   d(z12)/d(t) = (1/(1-X)^6)*(exp(t/(1+0.067*t)))*A1*((exp(0.1*m)))*m
    1*(cos (p*m))*z^n
3   d(z21)/d(t) = (1/(1-X)^6)*(exp(t/(1+0.067*t)))*(A2)*(z^s)*((exp
    (0.1*m)))*m1*(sin (p*m))*z^n
4   d(z22)/d(t) = (1/(1-X)^6)*(exp(t/(1+0.067*t)))*(A2)*(z^s)*((exp
    (0.1*m)))*m1*(cos (p*m))*z^n
5   d(z31)/d(t) = (1/(1-X)^2)*(1/k)*(exp(t/(1+0.067*t)))*A3*(z^s)*f3*
    ((exp(0.1*m2)))*m21*((sin (p*m2))^1)*z^n
6   d(z32)/d(t) = (1/(1-X)^2)*(1/k)*(exp(t/(1+0.067*t)))*A3*f3*(z^s)*
    ((exp(0.1*m2)))*m21*1*(cos (p*m2))*z^n
7   d(X1)/d(t) = m1*A*(z^2)/((1-X1)^6)
8   d(X2)/d(t) = m1*A0*(z^2)/((1-X2)^2)
```

Explicit equations

```
1   m = 0.01*t
2   m1 = 0.01
3   m2 = 0.0868*t - 0.0386*(t^2)+ 0.0080*(t^3) - 0.000756*(t^4) +
    2.6*(10^-5)*(t^5)
4   p = (10^2)
5   fi = 0.05
6   c1 = 3*(1-exp(-0.4*t))^0
7   c2 = 5*(1-exp(-0.4*t))^0
8   t2 = 6
9   t1 = 4
10  k = 10
11  n = 1.0
12  s = 1.0
13  E = (0.625-0.375*(tanh(c1*(t-t1))))*(1-fi)+(k)*(0.625-0.375*
    (tanh(c2*(t-t2))))*(fi)
14  z1 = (cos(p*m))*z11-(sin(p*m))*z12
15  z2 = (cos(p*m))*z21-(sin(p*m))*z22
16  Y1 = (0.15)*(sin (p*m))*E
17  A1 = 1
```

18 deff1 = (1+fi*((t+1)^0.333-1))^3 - 1

19 d = 1*t

20 A2 = 0.1*(exp(-0.4*t))*((1+2.5*fi+14.7*(fi^2)))*deff1

21 deff = (fi*d)/(1 + 0.333*(1 - fi)*d)

22 z3 = (cos(p*m2))*z31-(sin(p*m2))*z32

23 X = X1+X2

24 A = 2

25 A0 = 2

26 f3 = if (t<0) then (0) else (1)

27 f31 = (1/16)*t*(8-t)

28 f32 = (exp(-0.04*t))

29 f33 = (sin(3.14*t/4))^2

30 f34 = (1/64)*t^2

31 f35 = if (t<4) then (1) else (0)

32 f36 = if (t<4) then (0) else (1)

33 z = Y1-(z1+z2)*((exp(-0.1*m)))-z3*(exp(-0.1*m2))+0.175*0

34 m21 = 0.0868 - 0.0772*t + 0.024*(t^2) - 0.003*(t^3) + 1.3*(10^-4)* (t^4)

35 A3 = deff1

Model: $X = a1*t + a2*t^2 + a3*t^3 + a4*t^4 + a5*t^5$

Variable	Value
a1	0.1195218
a2	–0.1180537
a3	0.0372989
a4	0.0046664
a5	0.000201

$$X = \omega = 0.12\theta - 0.118\theta^2 + 0.0373\ \theta^3 - 0.00467\theta^4 + 0.0002\theta^5 \qquad (2.27)$$

Thus, it is clear that the criterion for fatigue of a composite is at the same time a condition for failure is the equality accepted here (as well as in many other cases for life prediction at high temperature).

$N_f = 1.75 = m_{max} p/2\pi) = 0.11(100)/2\pi = 1.75$ and $T_f = 1260[^\circ K](T_f = \beta T_* \theta_{max} + 600 = = 11(60) + 600 = 1260[^\circ K] = 960[^\circ C]; \beta T_* = 0.1 (600))$

Example 2.7—p = 100; a_0 = 0.25

Differential equations

1 d(z11)/d(t) = (1/(1-X)^6)*(exp(t/(1+0.067*t)))*A1*((exp(0.1*m)))*m1 *(sin (p*m))*z^n

2 d(z12)/d(t) = (1/(1-X)^6)*(exp(t/(1+0.067*t)))*A1*((exp(0.1*m)))*m1 *(cos (p*m))*z^n

3 d(z21)/d(t) = (1/(1-X)^6)*(exp(t/(1+0.067*t)))*(A2)*(z^s)*((exp (0.1*m)))*m1*(sin (p*m))*z^n

4 d(z22)/d(t) = (1/(1-X)^6)*(exp(t/(1+0.067*t)))*(A2)*(z^s)*((exp (0.1*m)))*m1*(cos (p*m))*z^n

5 d(z31)/d(t) = (1/(1-X)^2)*(1/k)*(exp(t/(1+0.067*t)))*A3*(z^s)*f3* ((exp(0.1*m2)))*m21*((sin (p*m2))^1)*z^n

6 d(z31)/d(t) = (1/(1- X)^2)*(1/k)*(exp(t/(1+0.067*t)))*A3*(z^s)*f3*((exp(0.1*m2)))*m21* ((cos(p*m2))*z^n

7 d(X1)/d(t) = m1*A*(z^2)/((1-X1)^6)

8 d(X2)/d(t) = m1*A0*(z^2)/((1-X2)^2)

Model: X = a1*t + a2*t^2 + a3*t^3 + a4*t^4 + a5*t^5

Variable	Value
a1	0.1008272
a2	–0.1052083
a3	0.035789
a4	–0.0048243
a5	0.0002239

$$X = \omega = 0.1\theta - 0.105\theta^2 + 0.0358\,\theta^3 - 0.00482\theta^4 + 0.000224\theta^5 \quad (2.28)$$

$N_f = 1.59 = m_{max(}p/2\pi) = 0.1(100)/2\pi = 1.59$ and $T_f = 672[^0K]$ $(T_f = \beta T_* \theta_{max} + 600 =$ $= 10(60) + 600 = 1200[^0K] = 900[^0C]$; $\beta T_* = 0.1\ (600))$

Example 2.3—p = 100; a_0 = 0.35

Differential equations

1 d(z11)/d(t) = (1/(1-X)^6)*(exp(t/(1+0.067*t)))*A1*((exp(0.1*m)))*m1 *(sin (p*m))*z^n

2 d(z12)/d(t) = (1/(1-X)^6)*(exp(t/(1+0.067*t)))*A1*((exp(0.1*m)))*m1 *(cos (p*m))*z^n

3 d(z21)/d(t) = (1/(1-X)^6)*(exp(t/(1+0.067*t)))*(A2)*(z^s)*((exp(0.1* m)))*m1*(sin (p*m))*z^n

4 d(z22)/d(t) = (1/(1-X)^6)*(exp(t/(1+0.067*t)))*(A2)*(z^s)*((exp(0.1* m)))*m1*(cos (p*m))*z^n

5 d(z31)/d(t) = (1/(1-X)^2)*(1/k)*(exp(t/(1+0.067*t)))*A3*(z^s)*f3*((ex p(0.1*m2)))*m21*((sin (p*m2))^1)*z^n

6 d(z32)/d(t) = (1/(1-X)^2)*(1/k)*(exp(t/(1+0.067*t)))*A3*f3*(z^s)*((ex p(0.1*m2)))*m21*1*(cos (p*m2))*z^n

7 d(X1)/d(t) = m1*A*(z^2)/((1-X1)^6)

8 d(X2)/d(t) = m1*A0*(z^2)/((1-X2)^2)

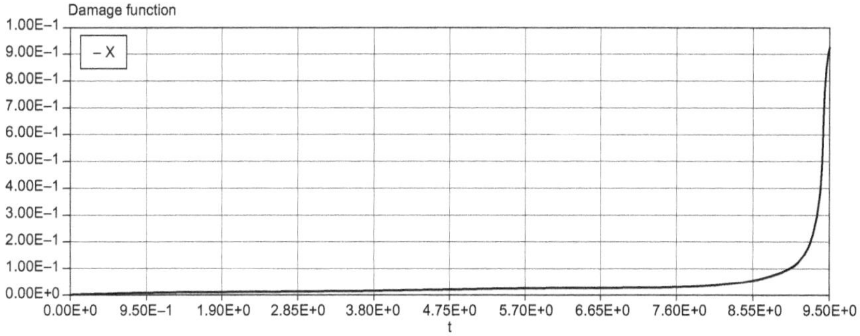

Figure 2.5. Damage function $a_0 = 0.45$.

Model: X = a1*t + a2*t^2 + a3*t^3 + a4*t^4 + a5*t^5

Variable	Value
a1	0.1132589
a2	–0.1223131
a3	0.0440037
a4	–0.0062786
a5	0.0003084

$$X = \omega = 0.113\theta - 0.122\theta^2 + 0.044\theta^3 - 0.0063\theta^4 + 0.00031\theta^5 \quad (2.29)$$

$N_f = 1.502 = m_{max}(p/2\pi) = 0.0944(100)/2\pi = 1.502$ and $T_f = 672[^0K]$
$(T_f = \beta T_* \theta_{max} + 600 = 9.4(60) + 600 = 1164[^0K] = 864[^0C]; \beta T_* = 0.1\ (600))$

Example 2.4—p = 100 $a_0 = 0.45$

Differential equations

1 d(z11)/d(t) = (1/(1-X)^6)*(exp(t/(1+0.067*t)))*A1*((exp(0.1*m)))*m1 *(sin (p*m))*z^n
2 d(z12)/d(t) = (1/(1-X)^6)*(exp(t/(1+0.067*t)))*A1*((exp(0.1*m)))*m1 *(cos (p*m))*z^n
3 d(z21)/d(t) = (1/(1-X)^6)*(exp(t/(1+0.067*t)))*(A2)*(z^s)*((exp(0.1* m)))*m1*(sin (p*m))*z^n
4 d(z22)/d(t) = (1/(1-X)^6)*(exp(t/(1+0.067*t)))*(A2)*(z^s)*((exp(0.1* m)))*m1*(cos (p*m))*z^n
5 d(z31)/d(t) = (1/(1-X)^2)*(1/k)*(exp(t/(1+0.067*t)))*A3*(z^s)*f3*((ex p(0.1*m2)))*m21*((sin (p*m2))^1)*z^n
6 d(z32)/d(t) = (1/(1-X)^2)*(1/k)*(exp(t/(1+0.067*t)))*A3*f3*(z^s)*((ex p(0.1*m2)))*m21*1*(cos (p*m2))*z^n
7 d(X1)/d(t) = m1*A*(z^2)/((1-X1)^6)
8 d(X2)/d(t) = m1*A0*(z^2)/((1-X2)^2)

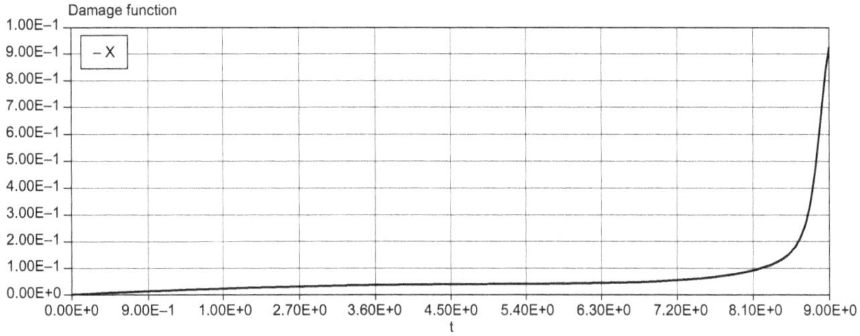

Figure 2.6. Damage function $a_0 = 0.45$.

Model: X = a1*t + a2*t^2 + a3*t^3 + a4*t^4

Variable	Value
a1	–0.0791857
a2	0.0704717
a3	–0.0161794
a4	0.0011103

$$X = \omega = -0.079\theta + 0.070\theta^2 - 0.016\ \theta^3 + 0.00111\theta^4 \quad (2.30)$$

$N_f = 1.41 = m_{max\,(}p/2\pi) = 0.0884(100)/2\pi = 1.41$ and $T_f = 1140[^0K]$
($T_f = \beta T_*\ \theta_{max} + 600 = 9.0(60) + 600 = 1140[^0K] = 840[^0C]$; $\beta T_* = 0.1\ (600)$)

Example 2.7—p = 100 $a_0 = 0.65$

Differential equations

1 d(z11)/d(t) = (1/(1-X)^6)*(exp(t/(1+0.067*t)))*A1*((exp(0.1*m)))*m1 *(sin (p*m))*z^n

2 d(z12)/d(t) = (1/(1-X)^6)*(exp(t/(1+0.067*t)))*A1*((exp(0.1*m)))*m1 *(cos (p*m))*z^n

3 d(z21)/d(t) = (1/(1-X)^6)*(exp(t/(1+0.067*t)))*(A2)*(z^s)*((exp(0.1* m)))*m1*(sin (p*m))*z^n

4 d(z22)/d(t) = (1/(1-X)^6)*(exp(t/(1+0.067*t)))*(A2)*(z^s)*((exp(0.1* m)))*m1*(cos (p*m))*z^n

5 d(z31)/d(t) = (1/(1-X)^2)*(1/k)*(exp(t/(1+0.067*t)))*A3*(z^s)*f3*((e xp(0.1*m2)))*m21*((sin (p*m2))^1)*z^n

6 d(z32)/d(t) = (1/(1-X)^2)*(1/k)*(exp(t/(1+0.067*t)))*A3*f3*(z^s)*((e xp(0.1*m2)))*m21*1*(cos (p*m2))*z^n

7 d(X1)/d(t) = m1*A*(z^2)/((1-X1)^6)

8 d(X2)/d(t) = m1*A0*(z^2)/((1-X2)^2)

Model: X = a1*t + a2*t^2 + a3*t^3 + a4*t^4

Variable	Value
a1	–0.0820526
a2	0.0819986
a3	–0.0201884
a4	0.0014843

$$X = \omega = -0.082\theta + 0.081\theta^2 - 0.020\ \theta^3 + 0.00148\theta^4 \quad (2.31)$$

$N_f = 1.3 = m_{max\,(}p/2\pi) = 0.082(100)/2\pi = 1.30$ and $T_f = 672[^0K]$
($T_f = \beta T_*\ \theta_{max} + 600 = 8.5(60) + 600 = 1110[^0K] = 810[^0C]$; $\beta T_* = 0.1\ (600)$)

Example 2.9—p = 100 $a_0 = 0.75$

Differential equations

1 d(z11)/d(t) = (1/(1-X)^6)*(exp(t/(1+0.067*t)))*A1*((exp(0.1*m)))*m1 *(sin (p*m))*z^n

2 d(z12)/d(t) = (1/(1-X)^6)*(exp(t/(1+0.067*t)))*A1*((exp(0.1*m)))*m1 *(cos (p*m))*z^n

3 d(z21)/d(t) = (1/(1-X)^6)*(exp(t/(1+0.067*t)))*(A2)*(z^s)*((exp(0.1* m)))*m1*(sin (p*m))*z^n

4 d(z22)/d(t) = (1/(1-X)^6)*(exp(t/(1+0.067*t)))*(A2)*(z^s)*((exp(0.1* m)))*m1*(cos (p*m))*z^n

5 d(z31)/d(t) = (1/(1-X)^2)*(1/k)*(exp(t/(1+0.067*t)))*A3*(z^s)*f3*((ex p(0.1*m2)))*m21*((sin (p*m2))^1)*z^n

6 d(z32)/d(t) = (1/(1-X)^2)*(1/k)*(exp(t/(1+0.067*t)))*A3*f3*(z^s)*((ex p(0.1*m2)))*m21*1*(cos (p*m2))*z^n

7 d(X1)/d(t) = m1*A*(z^2)/((1-X1)^6)

8 d(X2)/d(t) = m1*A0*(z^2)/((1-X2)^2)

Model: X = a1*t + a2*t^2 + a3*t^3 + a4*t^4

Variable	Value
a1	–0.0880671
a2	0.1143094
a3	–0.0327903
a4	0.0028061

$$X = \omega = -\,0.088\theta + 0.114\theta^2 - 0.0327\,\theta^3 + 0.0028\theta^4 \qquad (2.32)$$

$N_f = 1.11 = m_{max}$ (p/2π) = 0.07(100)/2π = 1.11 and $T_f = 672[^0K]$
($T_f = \beta T_* \,\theta_{max} + 600$ = 7.2(60) + 600 = 1032[^{0}K] = 732[^{0}C]; $\beta T_* = 0.1$ (600))

Example 2.10—p = 100 $a_0 = 0.95$

Differential equations

1 d(z11)/d(t) = (1/(1-X)^6)*(exp(t/(1+0.067*t)))*A1*((exp(0.1*m)))*m1 *(sin (p*m))*z^n

2 d(z12)/d(t) = (1/(1-X)^6)*(exp(t/(1+0.067*t)))*A1*((exp(0.1*m)))*m1 *(cos (p*m))*z^n

3 d(z21)/d(t) = (1/(1-X)^6)*(exp(t/(1+0.067*t)))*(A2)*(z^s)*((exp(0.1* m)))*m1*(sin (p*m))*z^n

4 d(z22)/d(t) = (1/(1-X)^6)*(exp(t/(1+0.067*t)))*(A2)*(z^s)*((exp(0.1*m)))*m1*(cos (p*m))*z^n

5 d(z31)/d(t) = (1/(1-X)^2)*(1/k)*(exp(t/(1+0.067*t)))*A3*(z^s)*f3*((exp(0.1*m2)))*m21*((sin (p*m2))^1)*z^n

6 d(z32)/d(t) = (1/(1-X)^2)*(1/k)*(exp(t/(1+0.067*t)))*A3*f3*(z^s)*((exp(0.1*m2)))*m21*1*(cos (p*m2))*z^n

7 d(X1)/d(t) = m1*A*(z^2)/((1-X1)^6)

8 d(X2)/d(t) = m1*A0*(z^2)/((1-X2)^2)

Example 2.11—p = 100 a_0 = 0.95

Model: X = a1*t + a2*t^2 + a3*t^3 + a4*t^4

Variable	Value
a1	–0.1321434
a2	0.1959173
a3	–0.0625652
a4	0.0059968

$$X = \omega = -0.132\theta + 0.196\theta^2 - 0.0625\,\theta^3 + 0.006\theta^4 \quad (2.33)$$

$N_f = 0.955 = m_{max\,(}p/2\pi) = 0.06(100)/2\pi = 0.955$ and $T_f = 978[^0K]$
$(T_f = \beta T_* \,\theta_{max} + 600 = 6.3(60) + 600 = 978[^0K] = 678[^0C]; \beta T_* = 0.1$

All results from Examples 2.4–2.10 are presented in a tabulated form (see Table 2.1) and graphically (see Fig. 2.11) below:

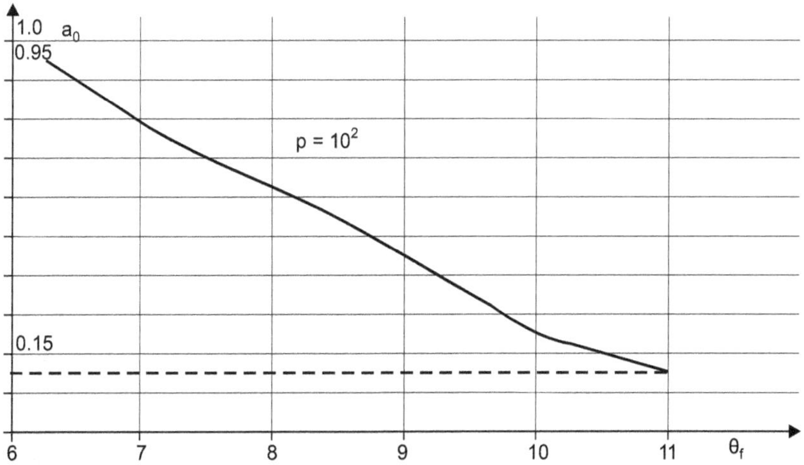

Figure 2.7. S – N Creep-fatigue Curve (p = 10^2 and $a_0 \in$ [0.15–0.95].

The Creep-fatigue Curve ($a_0 - N_f$) represents the classical form of fatigue of composite or nanocomposite materials mainly used based on experimental data. The failure criteria of composites in this case is based on a separate set of experimental data that describes behavior of that particular composite or nanocomposite. It becomes much more complex and costly then ordinary metal material. On the other hand the failure criteria based on an the assumption that composite and nanocomposite material fails if function f(ω) = 1 (damage factor is equal to 1) more reliable and at the same time analytically can be derived from the same tabulated data—see for example Tables (summary of results) and Graphs below.

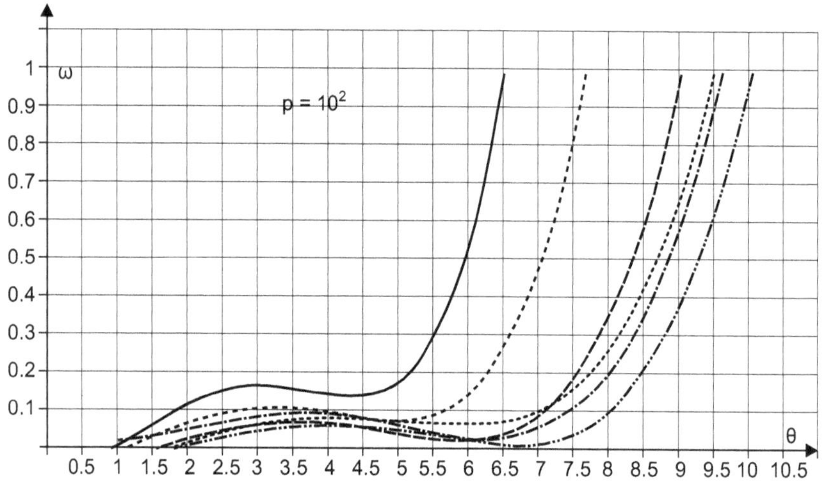

Figure 2.8. Damage Functions for $a_0 \in [0.15; 0.25, 0.95]$.

$$\omega(\theta) = -0.082*\theta + 0.081*\theta^2 - 0.020*\theta^3 + 0.00148*\theta^4 \quad (2.34)$$

All results from Examples 2.1–2.7 presented in a tabulated form (see Table 2.2) below:

Table 2.2. Summary of results: p = 100 t1 = θ_{gm} = 4; t2 = θ_{gf} = 6.

a/σ_{st}	0.15	0.25	0.35	0.45	0.55	0.75	0.95
p =100	θ_f = 11.0 ω = 1.0 m=0.11 N_f = 1.75 T_f = 960°C	θ_f =9.5 ω = 1.0 m=0.095 N_f = 1.5 T_f = 864°C	θ_f = 9.0 ω = 1.0 m=0.09 N_f =1.41 T_f = 840°C	θ_f = 8.5 ω = 1.0 m=0.085 N_f = 1.3 T_f = 810°C	θ_f = 7.2 ω = 1.0 m=0.072 N_f = 1.15 T_f = 732°C	θ_f =6.3 ω = 1.0 m=0.063 N_f = 0.965 T_f = 678°C	

When applying the CDM based cyclic creep-fatigue model to life prediction of composite and nanocomposite systems, the coupling between mechanical response and damage accumulation can be also obtained from Table 2.2, in other words the function ω (θ) can be constructed. The composite

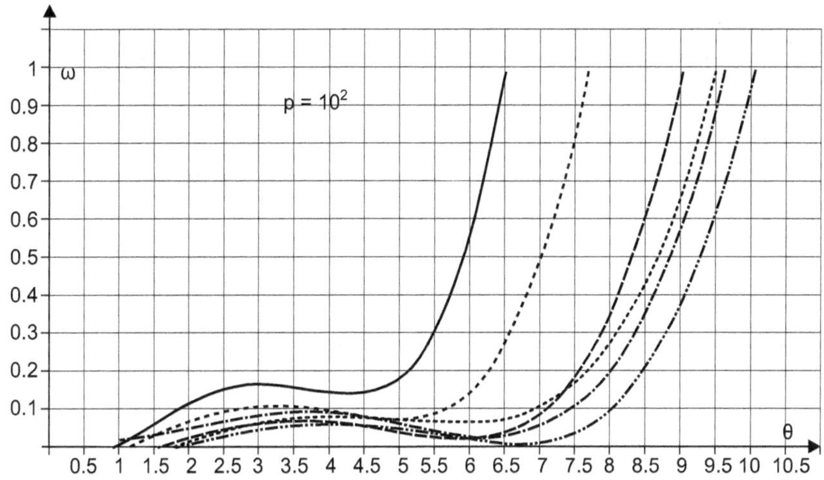

Figure 2.9. Damage Functions $\omega(\theta) = 1$ for $a_0 \in [0.15; 0.25 \ldots\ldots 0.95]$.

failure criteria remain the same: $\omega(\theta) = 1$. This type of presentation is very important whenever analyzing the creep-fatigue process from a probabilistic point of view. The detailed analysis of the probabilistic approach is presented in Chapter 5.

The results of fatigue-creep analysis in the case of different frequencies, p ($1 < p < 10^7$), in the nonlinear integral type Equation 2.20 (similar to the preceding analysis) are summarized and presented below in Table 2.2.

Example 2.9—p = 100 $a_0 = 0.75$

Differential equations

1 d(z11)/d(t) = (1/(1-X)^6)*(exp(t/(1+0.067*t)))*A1*((exp(0.1*m)))*m1 *(sin (p*m))*z^n

2 d(z12)/d(t) = (1/(1-X)^6)*(exp(t/(1+0.067*t)))*A1*((exp(0.1*m)))*m1 *(cos (p*m))*z^n

3 d(z21)/d(t) = (1/(1-X)^6)*(exp(t/(1+0.067*t)))*(A2)*(z^s)*((exp(0.1* m)))*m1*(sin (p*m))*z^n

4 d(z22)/d(t) = (1/(1-X)^6)*(exp(t/(1+0.067*t)))*(A2)*(z^s)*((exp(0.1* m)))*m1*(cos (p*m))*z^n

5 d(z31)/d(t) = (1/(1-X)^2)*(1/k)*(exp(t/(1+0.067*t)))*A3*(z^s)*f3*((e xp(0.1*m2)))*m21*((sin (p*m2))^1)*z^n

6 d(z32)/d(t) = (1/(1-X)^2)*(1/k)*(exp(t/(1+0.067*t)))*A3*f3*(z^s)*((e xp(0.1*m2)))*m21*1*(cos (p*m2))*z^n

7 d(X1)/d(t) = m1*A*(z^2)/((1-X1)^6)

8 d(X2)/d(t) = m21*A0*(z^2)/((1-X2)^2)

Table 2.3. Summary of results: p = 1000.

a/σ_{st}	0.15	0.25	0.35	0.45	0.55	0.75	0.95
p = 1000	$\theta_f = 14.0$ $\omega = 1.0$ m=0.14 $N_f = 22.3$ $T_f = 1140$	$\theta_f = 13.5$ $\omega = 1.0$ m=0.135 $N_f = 21.5$ $T_f = 111$	$\theta_f = 13.0$ $\omega = 1.0$ m=0.13 $N_f = 20.7$ $T_f = 1080$	$\theta_f = 12.0$ $\omega = 1.0$ m=0.12 $N_f = 19.1$ $T_f = 1020$	$\theta_f = 11.5$ $\omega = 1.0$ m=0.115 $N_f = 18.3$ $T_f = 990$	$\theta_f = 9.0$ $\omega = 1.0$ m=0.09 $N_f = 14.3$ $T_f = 840$	$\theta_f = 6.3$ $\omega = 1.0$ m=0.063 $N_f = 10.0$ $T_f = 678$

Example 2.9—p = 1000 t1 = $\boldsymbol{\theta}_{gm} = 4$; t2 = $\boldsymbol{\theta}_{gf} = 6$; m = 0.01$\boldsymbol{\theta}$; m1 = 0.01

Model: y = a0 + a1*x + a2*x^2 + a3*x^3 + a4*x^4

Variable	Value
a0	0.7068373
a1	0.2673268
a2	–0.061852
a3	0.0050385
a4	–0.0001562

$$y = a_0 = S = 0.707 + 0.267(\theta_f) - 0.618(\theta_f)^2 + 0.005(\theta_f)^3 - 0.00016\,(\theta_f)^4 \quad (2.35)$$

Example 2.10—p = 10000; $N_f = 0.01\theta_f(p/2\pi)$

Table 2.3 Summary of results: p = 10000

For example: if $a_0 = 0.15$ – is given, then: $N_f = m_{max\,(}p/2\pi) = 0.14(1000)/2\pi = 22.3$ and $T_f = \beta T_* \theta_{max} + 600 = 14(60) + 600 = 1440[^0K] = 1140[^0C]$; $\beta T_* = 0.1$

If $a_0 = 0.45$ – is given, then: $N_f = m_{max\,(}p/2\pi) \approx 0.12(1000)/2\pi = 19.1$ and $T_f = \beta T_* \theta_{max} + 600 = 12(60) + 600 = 1320[^0K] = 1020[^0C]$

Model: y = a1*x + a2*x^2 + a3*x^3 + a4*x^4

Variable	Value
a1	0.3549026
a2	–0.0423577
a3	0.0019918
a4	–3.553E-05

Example 2.11—p = 10000

Table 2.4. Summary of results: p = 10000.

a/σ_{st}	0.15	0.25	0.35	0.45	0.55	0.75	0.95
p = 10^4	$\theta_f = 19.7$ $\omega = 1.0$ m=0.197 $N_f = 313.5$ $T_f =$ 1482°C	$\theta_f = 17.7$ $\omega = 1.0$ m=0.177 $N_f =$ 281.7 $T_f =$ 1362°C	$\theta_f = 16.7$ $\omega = 1.0$ m=0.167 $N_f = 265.8$ $T_f =$ 1302°C	$\theta_f = 15.7$ $\omega = 1.0$ m=0.157 $N_f = 250.0$ $T_f =$ 1242°C	$\theta_f = 14.3$ $\omega = 1.0$ m=0.143 $N_f =$ 227.6 $T_f =$ 1158°C	$\theta_f = 12.3$ $\omega = 1.0$ m=0.123 $N_f =$ 195.8 $T_f =$ 1038°C	$\theta_f = 9.3$ $\omega = 1.0$ m=0.093 $N_f = 148$ $T_f =$ 858°C

$$y = \omega = -0.0935\,a_0 + 1.565(a_0)^2 - 3.4(a_0)^3 + 2.64(a_0)^4 \quad (2.36)$$

Example 2.12—p = 100000 t1 = $\theta_{gm} = 4$; t2 = $\theta_{gf} = 6$

Table 2.5. Summary of results: p = 100000.

a/σ_{st}	0.15	0.25	0.35	0.45	0.55	0.75	0.95
p = 10^5	$\theta_f = 20.0$ $\omega = 1.0$ m=0.2 $N_f = 3183$ $T_f =$ 1500°C	$\theta_f = 18.5$ $\omega = 1.0$ m=0.185 $N_f = 2944$ $T_f =$ 1410°C	$\theta_f = 17.5$ $\omega = 1.0$ m=0.175 $N_f = 2785$ $T_f =$ 1350°C	$\theta_f = 16.5$ $\omega = 1.0$ m=0.165 $N_f = 2626$ $T_f =$ 1290°C	$\theta_f = 15.5$ $\omega = 1.0$ m=0.155 $N_f = 2467$ $T_f =$ 1230°C	$\theta_f = 13.1$ $\omega = 1.0$ m=0.131 $N_f = 2085$ $T_f =$ 1086°C	$\theta_f = 10.7$ $\omega = 1.0$ m=0.107 $N_f = 1703$ $T_f =$ 942°C

$$y = \omega = 0.225\,a_0 - 2.763(a_0) + 12.156(a_0)^2 - 20.351(a_0)^3 + 12.32(a_0)^4 \quad (2.40)$$

Example 2.13—p = 1000000 t1 = $\theta_{gm} = 4$; t2 = $\theta_{gf} = 6$

Table 2.6. Summary of results: p = 1000000.

a/σ_{st}	0.15	0.25	0.35	0.45	0.55	0.75	0.95
p = 10^6	$\theta_f = 11.0$ $\omega = 1.0$ m=0.11 $N_f = 1.75$ $T_f = 960$°C	$\theta_f = 10.0$ $\omega = 1.0$ m=0.1 $N_f = 1.59$ $T_f = 900$°C	$\theta_f = 9.5$ $\omega = 1.0$ m=0.095 $N_f = 1.5$ $T_f = 864$°C	$\theta_f = 9.0$ $\omega = 1.0$ m=0.09 $N_f = 1.41$ $T_f = 840$°C	$\theta_f = 8.5$ $\omega = 1.0$ m=0.085 $N_f = 1.3$ $T_f = 810$°C	$\theta_f = 13.5$ $\omega = 1.0$ m=0.135 $N_f = 21486$ $T_f =$ 1110°C	$\theta_f = 11.5$ $\omega = 1.0$ m=0.115 $N_f = 18303$ $T_f = 990$°C

$$y = \omega = 0.191\,a_0 - 2.488(a_0) + 11.38(a_0)^2 - 19.463(a_0)^3 + 11.739(a_0)^4 \quad (2.41)$$

Example 2.14—p = 10000000 t1 = $\theta_{gm} = 4$; t2 = $\theta_{gf} = 6$

Table 2.7. Summary of results: p = 1000000.

a/σ_{st}	0.15	0.25	0.35	0.45	0.55	0.75	0.95
p = 10^7	θ_f = 11.0 ω = 1.0 m=0.11 N_f = 1.75 T_f = 960°C	θ_f =10.0 ω = 1.0 m=0.1 N_f = 1.59 T_f = 900°C	θ_f =9.5 ω = 1.0 m=0.095 N_f = 1.5 T_f = 864°C	θ_f = 9.0 ω = 1.0 m=0.09 N_f =1.41 T_f = 840°C	θ_f = 8.5 ω = 1.0 m=0.085 N_f = 1.3 T_f = 810°C	θ_f = 14 ω = 1.0 m=0.14 N_f =2.2(10^5) T_f =1140°C	θ_f =12.0 ω = 1.0 m=0.12 N_f =1.9(10^5) T_f =1020°C

$$y = \omega = 0.191\,a_0 - 2.488(a_0) + 11.38(a_0)^2 - 19.463(a_0)^3 + 11.739(a_0)^4 \qquad (2.42)$$

Summarizing results from Tables 2.1–2.7 and p = 1 till p = 10^7 (see Fig. 2.10):

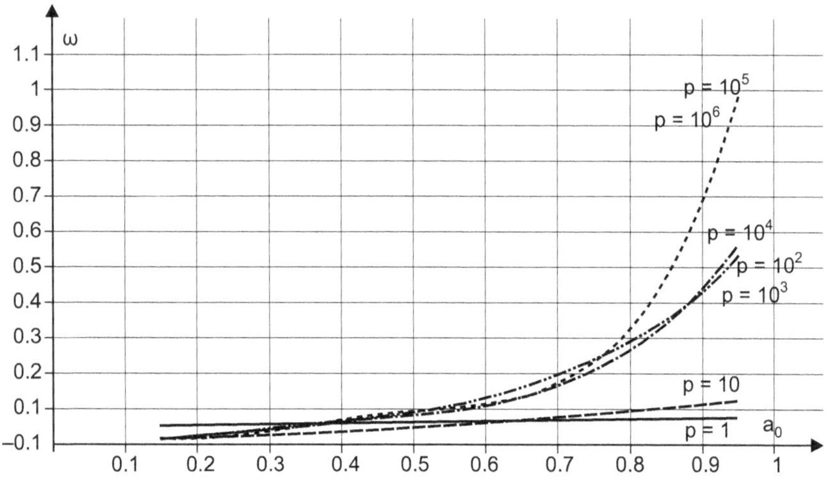

Figure 2.10. Summarizing results from Tables 2.1–2.7.

The current constitutive model has been formulated in a small strain framework, thus it can considered as an additive decomposition of the total strain tensor. This treatment can simplify the model structure and facilitate future numerical calculations. The Young's modulus E can be identified from the slope of the linear region of the stress-strain curves during the uniaxial loading tests. In the strategy of stress range separation, the initial yield stress k serves as the boundary stress between the low and high stress ranges. Within the framework of the current model, the value of k can be determined by using the uniaxial loading-dwell tests, which just equals the limiting value of the total stress after a long holding time.

Table 2.8. Summary of results: nonlinear fatigue-creep model ($1 < p < 10^7$).

a/σ_{st}	0.15	0.25	0.35	0.45	0.55	0.75	0.95
$X=\omega$ $p = 1$	$\omega = 0.0523$ $\theta = 10.25$ $\theta = 12.25$	$\omega = 0.055$ $\theta = 10.25$ $\theta = 12.25$	$\omega = 0.058$ $\theta = 10.25$ $\theta = 12.25$	$\omega = 0.060$ $\theta = 10.25$ $\theta = 12.25$	$\omega = 0.06$ $\theta = 10.2$ $\theta = 12.25$	$\omega = 0.070$ $\theta = 10.2$ $\theta = 12.25$	$\omega = 0.077$ $\theta = 10.25$ $\theta = 12.25$
$p = 10$	$\omega = 0.109$ $\theta = 10.25$	$\omega = 0.165$ $\theta = 10.25$	$\omega = 0.272$ $\theta = 10.25$	$\omega = 0.328$ $\theta = 9.55$	$\omega = 0.61$ $\theta = 6.25$	$\omega = 0.716$ $\theta = 6.00$	$\omega = 0.821$ $\theta = 4.20$
$X=\omega$ $p = 10^2$	$\omega = 0.0523$ @ $\theta = 9.25$ $\omega = 1.0$ @$\theta=10.55$	$\omega = 0.0671$ @ $\theta = 9.25$ $\omega = 1.0$ @$\theta=10.55$	$\omega =$ 0.0973 @ $\theta =$ 9.25 $\omega = 1.0$ @$\theta=10.55$	$\omega = 0.144$ @ $\theta = 9.25$ $\omega = 1.0$ @$\theta=8.25$	$\omega = 0.211$ @ $\theta = 9.25$ $\omega = 1.0$ @$\theta=6.95$	$\omega = 0.529$ @ $\theta =$ 3.25 $\omega = 1.0$ @$\theta=3.95$	$\omega = 0.891$ @ $\theta =$ 2.45 $\omega = 1.0$ @$\theta=2.45$
$X=\omega$ $p = 10^3$	$\omega = 0.0731$ @ $\theta = 9.25$ $\omega = 1.0$ @$\theta=10.55$	$\omega = 0.108$ @ $\theta = 9.25$ $\omega = 1.0$ @$\theta=10.55$	$\omega = 0.176$ @ $\theta =$ 9.25 $\omega = 1.0$ @$\theta=10.0$	$\omega = 0.324$ @ $\theta = 9.25$ $\omega = 1.0$ @$\theta=8.25$	$\omega = 0.492$ @ $\theta = 6.65$ $\omega = 1.0$ @$\theta=6.95$	$\omega = 0.623$ @ $\theta =$ 3.75 $\omega = 1.0$ @$\theta=3.95$	0.887 @$\theta=2.85$ $\omega = 1.0$ @$\theta=2.95$ $N_f =$
$X=\omega$ $p = 10^4$	$\omega = 0.0674$ @ $\theta = 9.75$ $\omega = 1.0$ @$\theta=10.65$	$\omega = 0.103$ @ $\theta = 9.75$ $\omega = 1.0$ @$\theta=10.65$	$\omega = 0.163$ @ $\theta =$ 9.75 $\omega = 1.0$ @$\theta=10.65$	$\omega = 0.268$ @ $\theta = 9.75$ $\omega = 1.0$ @$\theta=12.65$	$\omega = 0.368$ @ $\theta = 6.55$ $\omega = 1.0$ @$\theta=6.65$	$\omega = 0.570$ @ $\theta =$ 3.75 $\omega = 1.0$ @$\theta=3.95$	$\omega = 0.697$ @$\theta=2.75$ $\omega = 1.0$ @$\theta=2.95$
$X=\omega$ $p = 10^5$	$\omega = 0.0656$ @ $\theta = 9.75$ $\omega = 1.0$ @$\theta=10.65$	$\omega = 0.103$ @ $\theta = 9.75$ $\omega = 1.0$ @$\theta=10.65$	$\omega = 0.163$ @ $\theta =$ 9.75 $\omega = 1.0$ @$\theta=10.65$	$\omega = 0.268$ @ $\theta = 9.75$ $\omega = 1.0$ @$\theta=12.65$	$\omega = 0.368$ @ $\theta = 6.55$ $\omega = 1.0$ @$\theta=6.65$	$\omega = 0.570$ @ $\theta =$ 3.75 $\omega = 1.0$ @$\theta=3.95$	0.705 @$\theta=2.75$ $\omega = 1.0$ @$\theta=2.95$
$X=\omega$ $p = 10^6$	$\omega = 0.0656$ @ $\theta = 9.75$ $\omega = 1.0$ @$\theta=10.65$	$\omega = 0.103$ @ $\theta = 9.75$ $\omega = 1.0$ @$\theta=10.65$	$\omega = 0.163$ @ $\theta =$ 9.75 $\omega = 1.0$ @$\theta=10.65$	$\omega = 0.268$ @ $\theta = 9.75$ $\omega = 1.0$ @$\theta=12.65$	$\omega = 0.368$ @ $\theta = 6.55$ $\omega = 1.0$ @$\theta=6.65$	$\omega = 0.570$ @ $\theta =$ 3.75 $\omega = 1.0$ @$\theta=3.95$	0.703 @$\theta=2.75$ $\omega = 1.0$ @$\theta=2.95$ $N_f =$ 1,400,000
$X=\omega$ $p = 10^7$ $\omega = 1.0$	0.0656 @ $\theta = 9.75$ $\omega = 1.0$ @$\theta=10.65$	$\omega = 0.103$ @ $\theta = 9.75$ $\omega = 1.0$ @$\theta=10.65$	$\omega = 0.163$ @ $\theta =$ 9.75 $\omega = 1.0$ @$\theta=10.65$	$\omega = 0.268$ @ $\theta = 9.75$ $\omega = 1.0$ @$\theta=12.65$	$\omega = 0.368$ @ $\theta = 6.55$ $\omega = 1.0$ @$\theta=6.65$	$\omega = 0.57$ @ $\theta =$ 3.75 $\omega = 1.0$ @$\theta=3.95$	$\omega = 0.703$ @$\theta=2.75$ $\omega = 1.0$ @$\theta=2.95$

Instead of the standard uniaxial stress $\sigma = F/S$, it is convenient to introduce the effective stress for the damaged material,

$$\bar{\sigma} = \frac{F}{S - S_\omega} = \frac{F}{S(1 - S_\omega / S)} = \frac{\sigma}{1 - \omega} \tag{2.43}$$

Associated to a strain equivalence principle, the 1D effective stress σ is related to the elastic strain of the material by the uniaxial Hooke's law: $\bar{\sigma} = E\varepsilon$, where E is the elastic modulus of elasticity of undamaged material. It follows that the constitutive law for the standard stress σ takes the form:

$$\sigma = (1 - \omega)E\varepsilon \tag{2.44}$$

For the uniaxial model formulation, equation [4] must be completed by the damage evolution law which can be considered in the form of dependence between the damage variable ω and the applied load:

$$\omega = g(\varepsilon) \tag{2.45}$$

A priori, the function g can be identified from uniaxial tension test. It must be noted that the relation between ω and ε is valid only in the monotonous external loading regime. In the unloading and reloading phase, the damage variable kept its maximum value reached before. A classical way to describe in a unified manner these different loading regimes consists in introducing a variable κ which characterizes the maximum level of strain reached in the material before the current time t: $\kappa(t) = \max \varepsilon(\tau)$ for $\tau \leq t$. The damage evolution relation [5] can then be recast in form of: $\omega = g(k)$, which remains valid for any kind of loading regime.

References

[1] Richard Schapery. A method for mechanical state characterization of inelastic composite laminates with damage. Conference: Proceedings, 7th International Conference on Fracture, Houston, 1989.
[2] Fung Y. C. Foundations of Solid Mechanics (Prentice-Hall, Englewood Cliffs, NJ, 1976).
[3] Malvern L. E. Introduction to the Mechanics of a Continuous Medium (Prentice-Hall, Englewood Cliffs, N J, 1969).
[4] Lemaitre J. and Desmorat, R. Engineering Damage Mechanics - Ductile, Creep, Fatigue and Brittle Failures. Springer, Berlin, 2007.
[5] Claude Bathias. An engineering point of view about fatigue of polymer matrix composite materials. International Journal of Fatigue 28: 1094–1099, France, 2006.
[6] Harris B. Fatigue in composite. CRC/WP; vol. 3; 2003.
[7] Degrieck J. and Paepegem W. V. Fatigue damage modeling of fiber-reinforced composite materials. Rev. Appl. Mech. Rev., 54: 279–300, 2004.
[8] Anastasia Muliana. Nonlinear viscoelastic-degradation model for polymeric based materials. International Journal of Solids and Structures, 51(1): 122–132, 2014.
[9] Kachanov L. M. Izvestiaya Academe Nauk SSR, Otdelenie Tekhnicheskii Nauk, 8: 26–31, 1958.
[10] Sullivan R. W. Development of a viscoelastic continuum damage model for cyclic loadin. Department of Aerospace Engineering, Mississippi State University, 2008.

[11] Razdolsky L. Fatigue-Creep Phenomenological Models of Composites and Nanocomposites. AIAA Propulsion and Energy Forum, AIAA 2019.
[12] Razdolsky L. Phenomenological Creep Models of Composites and Nanomaterials Deterministic and Probabilistic Approach. CRC Press Taylor & Francis Group, Boca Raton, FL, 2019.
[13] D'Amore A. and Grassia L. Phenomenological approach to the study of Hierarchical damage mechanisms in composite materials subjected to fatigue loadings. Composites Structures, 175, 1–6, 2017.
[14] D'Amore A. and Grassia L. Constitutive law describing the strength degradation kinetics of fiber-reinforced composites subjected to constant amplitude cyclic loading. Journal of Mechanics of Time-Dependent Materials, 20: 1–12, 2016.
[15] Ringel M., Roos E., Maile K. et al. Constitutive equations of adapted complexity for high temperature loading. pp. 638–648. *In*: ECCC Creep Conference, London, UK, 12–14 September, 2005.
[16] Vassilopoulos A. P. and Keller T. Fatigue of Fiber-Reinforced Composites; Springer: London, UK, 2013.
[17] Joris Degrieck and Wim Van Paepegem. Fatigue Damage Modeling of Fiber-reinforced Composite Materials: Review, Applied Mechanics Reviews, 54(4): 279–300, Belgium, 2001.
[18] Ruben Dirk Sevenois, Wim Van Paepegem. Fatigue Damage Modeling Techniques for Textile Composites: Review and Comparison with UD Composite Modeling Techniques, Applied Mechanics Reviews 67(2), 2015.
[19] Degrieck J. and Van Paepegem, W. Fatigue Damage Modeling of Fiber-Reinforced Composite Materials: Review. Applied Mechanics Reviews, 54(4): 279–300, 2001.
[20] Degrieck J. and Wim Van Paepegem. Fatigue Damage Modeling of Fiber-reinforced Composite Materials: Review. Applied Mechanics Reviews 54(4), 2001.
[21] Whitney J. M. and Nuismer R. J. Stress fracture criteria for laminated composite containing stress concentration. Journal Composite Materials, 8: 253–6, 1974.
[22] Ladeve`ze P. and Lubineau G. A computational meso-damage model for life prediction for laminates. CRC/WP; 3: 432–41, 2003.
[23] Bathias C. Fracture and fatigue of high performance composite materials: Mechanisms and prediction. Engineering Fractures Mechanic, 40:757–83, 1991.
[24] Krajcinovic D. Damage Mechanics (Elsevier, Amsterdam, 1996).
[25] Suresh S. Fatigue of Materials (Cambridge University Press, Cambridge, England, 2006.
[26] Marigo J. J. Modeling of brittle and fatigue damage for elastic material by growth of micro-voids. Engineering Fractures Mechanics, 21(4): 861–874, 1985.
[27] Gamstedt K. and Andersen S. I. S. Fatigue degradation and failure of rotating composite structures. Materials characterization and underlying mechanisms. Risø National Laboratory. Denmark. Forskningscenter Risoe. Risoe-R, 2001.
[28] Rouchon J. Fatigue and Damage Tolerance Evaluation of Structures: The Composite Materials Response. National Luchten Ruimtevaart laboratorium National Aerospace Laboratory NLR, 2009.
[29] Razdolsky L. Structural Fire Loads: Theory and Principles. McGraw—Hill Co. N.Y., N.Y., 2012.
[30] Prout E. G. and Tompkins F.C. The thermal decomposition of potassium permanganate. Trans. Faraday Soc., 40: 488–498, 1944.
[31] Razdolsky L. Fatigue-Creep Phenomenological Models of Composites and Nanocomposites. AIAA Propulsion and Energy Forum, AIAA, 2019.
[32] Razdolsky L. Phenomenological Creep Models of Composites and Nanomaterials, Boca Raton CRC Press, 2019.

[33] Croman R. B. Tensile Fatigue Performance of Thermoplastic Resin Composites Reinforced with Ordered Kevlar® Aramid Staple. pp. 572–577. *In*: Wu Y., Gu Z. and Wu R. (eds.). Proceedings of the Seventh International Conference on Composite Materials, 2, International Academic Publishers, Oxford, 1989.
[34] Razdolsky L. Probability Based High Temperature Engineering Springer Nature Publishing Co. AG Switzerland, 2017.
[35] Harris B. Fatigue in composite, CRC/WP; vol. 3, 2003.
[36] Razdolsky L. Fatigue-Creep Phenomenological Models of Composites and Nanocomposites. AIAA Propulsion and Energy Forum, AIAA, 2019.
[37] POLYMATH software.

Chapter 3

Phenomenological Creep-Fatigue Models

3.1 Introduction

This chapter proposes a phenomenological, physically-based, inelastic model to predict cyclic creep-fatigue taking into account accumulated damage under high temperature stress conditions in composite materials. A technique for identifying model parameters based on the minimum required set of experimental data has been developed. The phenomenological and adjacent thermodynamic [1,2] approaches for describing inelastic rheological deformation and failure of composites and nanocomposites under unsteady temperature loading give good results, and the expediency of their use in computational practice is beyond doubt. The aim of this work is to generalize the integral approach of a non-stationary creep process [3] to describe a class of phenomena occurring in a composite material under the combined action of static σ_{st} and cyclic loads with the amplitude value of the cyclic component of dimensionless stresses σ_a and frequency p. The restriction here is to consider the so-called multi-cycle loading at a frequency $p > 10$ Hz and an amplitude factor $\sigma_0 = \sigma_{st}/\sigma_a$ not exceeding a certain critical value $\sigma_0 = 0.95$, that is, $\sigma_0 \in [0.15, 0.95]$. Thus, the integral type of cyclic creep and fatigue constitutive equation combined with the continuum damage mechanics in this case has the form:

$$E(\theta)[\theta]=\sigma(\theta)+\int_0^{\theta} e^{\frac{\tau}{1+\beta\tau}}K_1(\theta,\tau)\sigma^n(\tau)\,m\mathbf{1}d\tau+\int_0^{\theta} A_2 f_2[\sigma(\tau)]K_2(\theta,\tau)\sigma^n(\tau)\,m\mathbf{1}d\tau+$$

$$+\int_0^{\theta}\frac{E_0}{E_1}A_3 f_3[d(\tau)]K_3(\theta,\tau)\sigma(\tau)\,m\mathbf{21}d\tau$$

$$K_1(\theta,\tau)=\left(\frac{\mathbf{1}}{(\mathbf{1}-X)^6}\right)m\mathbf{1}(\tau)\sum_{i=1}^{N}\exp(-\alpha_i\,m(\theta))\exp(\alpha_i\,m(\tau))\left[\sin p[m(\theta)-m(\tau)]\right]$$

$$K_2(\theta,\tau)=e^{\frac{\tau}{1+\beta\tau}}\left(\frac{\mathbf{1}}{(\mathbf{1}-X)^6}\right)m\mathbf{1}(\tau)\sum_{i=1}^{N}\exp(-\beta_{i2}\,m(\theta))\exp(\beta_{i2}\,m(\tau))\left[\sin p[m(\theta)-m(\tau)]\right]$$

$$K_3(\theta,\tau)=e^{\frac{\tau}{1+\beta\tau}}\left(\frac{\mathbf{1}}{(\mathbf{1}-X)^6}\right)m\mathbf{21}\sum_{i=1}^{N}\exp(-\beta_{i3}\,m\mathbf{2}(\theta))\exp(\beta_{i3}\,m\mathbf{2}(\tau))\left[\sin p[m\mathbf{2}(\theta)-m\mathbf{2}(\tau)]\right]$$

$$f_2(\sigma)=\sigma^s;\;\; s=\mathbf{1},\mathbf{2},\mathbf{3}...,M;\;\; A_3=\text{deff}\mathbf{1}=\left(\mathbf{1}+fi*\left((t+\mathbf{1})\wedge\mathbf{0.333}-\mathbf{1}\right)\right)\wedge\mathbf{3}-\mathbf{1} \qquad (3.1)$$

$$d=a(\theta);\;\; k=E_0/E_1;$$

$$d(X\mathbf{1})/d(t)=m\mathbf{1}*A*(z\wedge\mathbf{2})/((\mathbf{1}-X\mathbf{1})\wedge\mathbf{6})$$

$$d(X\mathbf{2})/d(t)=m\mathbf{1}*A\mathbf{0}*(z\wedge\mathbf{2})/((\mathbf{1}-X\mathbf{2})\wedge\mathbf{2})$$

$$Y\mathbf{1}=E(\theta)[\theta]=\varepsilon_0[\sin(p\theta)]E;\;\; f\mathbf{3}=\text{if }(t<\mathbf{0})\text{ then}(\mathbf{0})\text{ else}(\mathbf{1});f\mathbf{31}=(\mathbf{1}/\mathbf{16})t(\mathbf{8}-t);$$

$$f\mathbf{32}=(\exp(-\mathbf{0.1}t));$$

$$f\mathbf{33}=(\sin(\mathbf{3.14}t/\mathbf{4}))^{\mathbf{2}};\;\; f\mathbf{34}=\text{if }(t<\mathbf{7.99})\text{ then }(\mathbf{0})\text{ else }(\mathbf{1});\;\; f\mathbf{35}=(\mathbf{1}/\mathbf{8})t$$

$$f\mathbf{36}=\text{if }(t<\mathbf{4})\text{ then }(\mathbf{1})\text{ else }(\mathbf{0});\;\; f\mathbf{37}=\text{if }(t<\mathbf{4})\text{ then}(\mathbf{0})\text{ else}(\mathbf{1})$$

$$f\mathbf{38}=\text{if }(t<\mathbf{0})\text{ then}(\mathbf{1})\text{ else}(\mathbf{0})$$

$$E=\left(\mathbf{0.625}-\mathbf{0.375}*\left(\tanh\left(\mathbf{5}*(t-\mathbf{3})\right)\right)\right)(\mathbf{1}-\varphi)+(\mathbf{1}/k)\left(\mathbf{0.625}-\mathbf{0.375}\left(\tanh\left(\mathbf{5}(t-\mathbf{5})\right)\right)\right)\varphi$$

$$fi=\varphi=\mathbf{0.05};k=\mathbf{0.1};\;\; n=s=\mathbf{1.0};$$

$$A2=\mathbf{0.1}*\left(\exp(-\mathbf{0.4}\theta)\right)\left(\left(\mathbf{1}+\mathbf{2.5}\varphi+\mathbf{14.7}\left(\varphi^{\mathbf{2}}\right)\right)\right)\text{deff}\mathbf{1}\wedge\mathbf{1}$$

Numerical analysis and results given below in the form of examples, graphs and tables do not require significant additional explanations. Let's briefly outline the method for determining the constants in the cyclic creep-fatigue law (3.1) and damage function ω parameters.

1. Sets of experimental results obtained at a given temperature and different stresses are used. For example, to determine the ratio of static stresses and vibration amplitude in Equation (3.1), it is also necessary to have a set of experiments on specimen creep-fatigue up to failure under three different loads. It is possible to use long-term strength curves to determine the constants included in Equation (3.1). Two experiments are needed to

obtain the constants, and the third one is needed to check the reliability of the constants found.

2. To find the material constants in the creep-fatigue law, two deformation values are used at certain points in time from the creep curves at two different stresses. Solving the differential Equation (3.1) and substituting the indicated values into it, the constants are determined.
3. To find the constants in the kinetic equation for the damage parameter, long-term strength curves are used. Solving the integral and differential Equations (3.1) together and substituting the value of the time to failure at two different locations are determined.
4. Using the found constants, the calculated curves of the dependence of deformation and parameters of the damage function on time, curves of long-term strength according to the equations of state for three values of stresses are constructed. Comparison of calculated curves and experimental data is made. If the relative error does not exceed the specified value (for example, 1.0%), then it was found that the material constants are considered correct. The conditions for the course of various technological processes, the loading of structural elements is often characterized by a cyclical nature of stress changes. The experimental derivation of the laws of cyclic loading and failure under uniaxial and stress states is a more difficult task in comparison with purely static experiments. This raises the problem of a theoretical description of cyclic creep-fatigue curves using the results of experiments carried out under static loading. The solution of this problem for the case of high-cycle loading is given in [4], for low-cycle loading-in [5]. The uniaxial fatigue analysis is used to estimate the fatigue life of a component under cyclic loading when the damage is initiated due to a uniaxial state of strain. The cyclic creep-fatigue life of a component refers to the fatigue life defined as the number of cycles (N_f).

The stress in a cycle can be described by stress amplitude (a_0) and static stress (a_{st}). Since a_0 is the primary factor affecting N_f, it is often chosen as the controlled or independent variable, and consequently, N_f is the dependent variable on a_0. To increase the reliability and durability of machines and structures operating in extreme thermal conditions, fundamental development of strength and durability is necessary in science [6]. This will allow us to obtain materials and alloys of especially high qualities with the maximum degree of physical (absence of defects of various structural levels), chemical (more uniform distribution of impurities throughout the volume with minimum concentration of impurities) and structural homogeneity. Nevertheless, failure to work or fatigue developments almost inevitably lead to irreversible consequences [7]. Therefore, an integral part of this problem is the achievement of a very high degree of quality of composites and nanocomposites. This is due to the fact that: firstly, there are reserves for improving the properties of composites through optimal design and heat treatment; secondly, it was found that some chemical elements can positively affect the durability of materials and structural elements. However, the existing imperfections of methods for assessing the strength of these materials often lead to the fact that large safety factors are accepted in the structural calculation methods. Therefore, products using composite materials are often designed with an excessive margin of safety, which reduces the efficiency of their use. These considerations fully relate to the methods for assessing the fatigue strength of composite materials. Before proceeding to a detailed analysis of the influence of individual dimensionless parameters on the behavior of composites under the action of heat flow, it is necessary to make sure that the proposed integral model (3.1) correctly (from the qualitative side) describes simple and physically obvious phenomena of cyclic creep and fatigue of a composite material. For example, in case of a sharp rise to high temperatures, the number of cycles until complete failure should be less than in the case of a slower increase in temperature.

Let's start here with the analysis of the effect of different types of temperature – time relationships (external) on the cyclic creep-fatigue process. This function is called here (by definition) a direct function (also called an identity function). That is a function that always returns the same value as its argument. It is denoted by f(x) = x. On the coordinate plane, the graph of the direct function is y = x. Two functions, f and g, are inverses of each other when f [g(x)] and g [f (x)] equal x. The inverse function is denoted by $1^{-1}(x)$. Simple case: $\boldsymbol{\theta} = a\boldsymbol{\tau}$ and therefore: $\tau = m = [1/a]\ \boldsymbol{\theta}$ and Y1 = (a_0)*(sin (p*m))*E – instantaneous stress. Y1 is a given cyclic function of the original (or dimensionless) time.

3.2 Effect of temperature – time relationships on creep-fatigue behavior of composites

It should be noted that, as before, the criterion for the fatigue of a composite or nanocomposite is a non-decreasing function ω (θ) or a non-increasing function a(N), that is, (S – N) fatigue curve. Moreover, in the first case, the fatigue life limit of the composite is the equation ω (θ) = 1, where θ is the maximum value of the dimensionless temperature θ_{max}; and in the second case, each value of the stress amplitude a_0 on the analytical fatigue curve (S – N) corresponds to the fatigue life limit of the composite or vice versa: the specified fatigue life limit of the composite corresponds to the maximum value of the stress amplitude. Moreover, these criteria for composite fatigue are valid for each value of the frequency p. Looking ahead, it should be noted that the first criterion for composite fatigue (the function ω (θ) could be considered as a random function of the dimensionless temperature θ) is the basis for formulating the problem of a probabilistic approach to the criterion of composite fatigue and, correspondingly, its fatigue life limit.

Example 3.1—Example 3.1 is an illustration of a simple physical fact (and, at the same time, one of the qualitative tests of the model of cyclic creep-fatigue of a composite proposed above): with an increase in the rate of temperature change with time, the fatigue life limit of the composite decreases.

Data: $\theta = a\boldsymbol{\tau}$; $\boldsymbol{\tau} = m = [1/a]\, \boldsymbol{\theta}$ and $\boldsymbol{\tau}' = m1 = [1/a]\, X \approx 1$

'a' is small: time – temperature rate process (a = 10) → m =0.1θ and m1 = 0.1

The differential equations are as follows:

1 d(z11)/d(t) = (1/(1-X)^6)*(exp(t/(1+0.067*t)))*A1*((exp(0.1*m)))*m1 *(sin (p*m))*z^n

2 d(z12)/d(t) = (1/(1-X)^6)*(exp(t/(1+0.067*t)))*A1*((exp(0.1*m)))*m1 *(cos (p*m))*z^n

3 d(z21)/d(t) = (1/(1-X)^6)*(exp(t/(1+0.067*t)))*(A2)*(z^s)*((exp (0.1*m)))*m1*(sin (p*m))*z^n

4 d(z22)/d(t) = (1/(1-X)^6)*(exp(t/(1+0.067*t)))*(A2)*(z^s)*((exp (0.1*m)))*m1*(cos (p*m))*z^n

5 d(z31)/d(t) = (1/(1-X)^2)*(1/k)*(exp(t/(1+0.067*t)))*A3*(z^s)*f3* ((exp(0.1*m2)))*m21*((sin (p*m2))^1)*z^n

6 d(z32)/d(t) = (1/(1-X)^2)*(1/k)*(exp(t/(1+0.067*t)))*A3*f3*(z^s)* ((exp(0.1*m2)))*m21*1*(cos (p*m2))*z^n

7 d(X1)/d(t) = m1*A*(z^2)/((1-X1)^6)

8 d(X2)/d(t) = m1*A_0*(z^2)/((1-X2)^2)

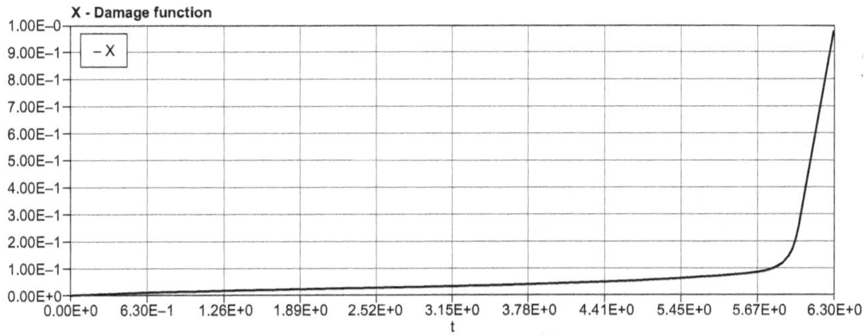

Figure 3.1. Damage function X.

Fatigue failure dimensionless temperature θ_f = 6.3 and τ = m= 0.63 → $N_f = (p\tau_f)/2\pi$ = 10.0 and if β = 0.1θ T_* = 60, then T_f= 978 [^{0} K] ≈ 678 [°C]. Comparing with the temperature rate τ = m = [1/100] θ (see Chapter 02) the maximum temperature at failure point (X = ω = 1) is θ_f =11.0 and N_f = 1.75 << 10.0 as expected.

Consider now a nonlinear case-inverse function time-temperature rate process (a = 100) → m =0.01θ^2 and m1 = 0.02θ.

Example 3.1a—Nonlinear case-direct function (also called an identity function) and inverse functions are as follows:

Data: $\theta = \sqrt{a\tau}$; $\theta^2 = a\tau$; m = [1/a]$\boldsymbol{\theta}^2$ and τ' = m1 = [2/a]θ

If 'a' is large: time – temperature rate process (a = 100) → m =0.01θ^2 and m1 = 0.02θ

Graphically functions 'm' and 'm1'are presented below (see Fig. 3.2).

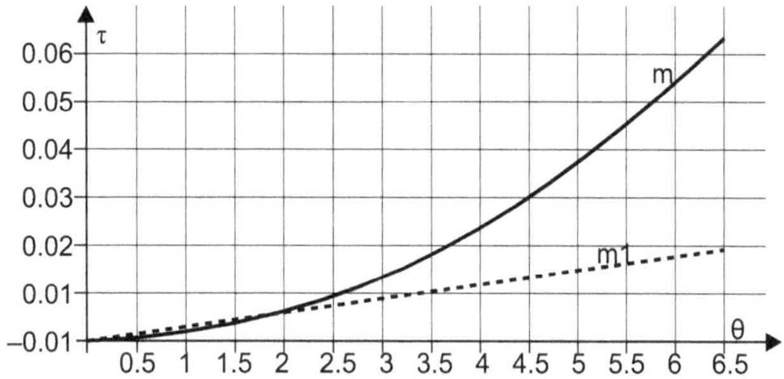

Figure 3.2. Reversed temperature – time function.

Differential equations

1 d(z11)/d(t) = (1/(1-X)^6)*(exp(t/(1+0.067*t)))*A1*((exp(0.1*m)))*m1*(sin (p*m))*z^n

2 d(z12)/d(t) = (1/(1-X)^6)*(exp(t/(1+0.067*t)))*A1*((exp(0.1*m)))*m1*(cos (p*m))*z^n

3 d(z21)/d(t) = (1/(1-X)^6)*(exp(t/(1+0.067*t)))*(A2)*(z^s)*((exp(0.1*m)))*m1*(sin (p*m))*z^n

4 d(z22)/d(t) = (1/(1-X)^6)*(exp(t/(1+0.067*t)))*(A2)*(z^s)*((exp(0.1*m)))*m1*(cos (p*m))*z^n

5 d(z31)/d(t) = (1/(1-X)^2)*(1/k)*(exp(t/(1+0.067*t)))*A3*(z^s)*f3*((exp(0.1*m2)))*m21*((sin (p*m2))^1)*z^n

6 d(z32)/d(t) = (1/(1-X)^2)*(1/k)*(exp(t/(1+0.067*t)))*A3*f3*(z^s)*((exp(0.1*m2)))*m21*1*(cos (p*m2))*z^n

7 d(X1)/d(t) = m1*A*(z^2)/((1-X1)^6)

8 d(X2)/d(t) = m1*A0*(z^2)/((1-X2)^2)

Explicit equations

1 m = 0.01*t^2

2 m1 = 0.02*t

3 m2 = 0.0868*t - 0.0386*(t^2)+ 0.0080*(t^3) - 0.000756*(t^4) + 2.6*(10^-5)*(t^5)

4 p = (10^2)

5 fi = 0.05

6 c1 = 3*(1-exp(-0.4*t))^0

7 c2 = 5*(1-exp(-0.4*t))^0

8 t2 = 6

9 t1 = 4

10 k = 10

11 n = 1.0

12 s = 1.0

13 E = (0.625-0.375*(tanh(c1*(t-t1))))*(1-fi)+(k)*(0.625-0.375*(tanh(c2*(t-t2))))*(fi)

14 z1 = (cos(p*m))*z11-(sin(p*m))*z12

15 z2 = (cos(p*m))*z21-(sin(p*m))*z22

16 Y1 = (0.15)*(sin (p*m))*E

17 A1 = 1

18 deff1 = (1+fi*((t+1)^0.333-1))^3 - 1

19 d = 1*t

20 A2 = 0.1*(exp(-0.4*t))*((1+2.5*fi+14.7*(fi^2)))*deff1^1

21 deff = (fi*d)/(1 + 0.333*(1 - fi)*d)

22 z3 = (cos(p*m2))*z31-(sin(p*m2))*z32

23 X = X1+X2

24 A = 2

25 A_0= 2

26 f3 = if (t<0) then (0) else (1)

27 f31 = (1/16)*t*(8-t)

28 f32 = (exp(-0.04*t))

29 f33 = (sin(3.14*t/4))^2

30 f34 = (1/64)*t^2

31 f35 = if (t<4) then (1) else (0)

```
f36 = if (t<4) then (0) else (1)
z = Y1-(z1+z2)*((exp(-0.1*m)))-z3*(exp(-0.1*m2))+0.175*0
m21 = 0.0868 - 0.0772*t + 0.024*(t^2) - 0.003*(t^3) + 1.3*(10^-4)*
(t^4)
A3 = deff1
```

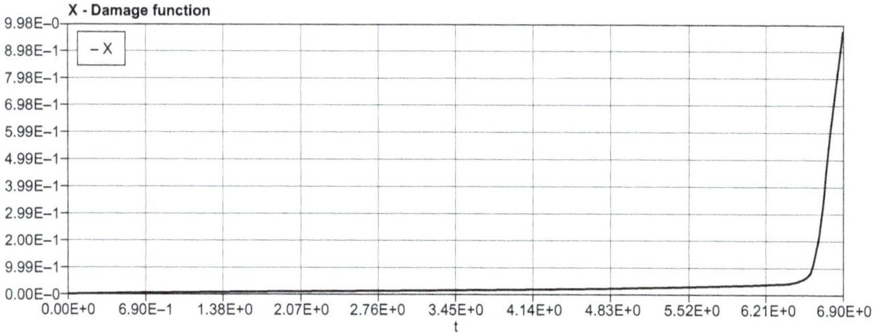

Figure 3.3. Damage function [case $\theta = \sqrt{a\tau}$].

Model: X = a1*t + a2*t^2 + a3*t^3 + a4*t^4 + a5*t^5

Variable	Value
a1	0.1778763
a2	–0.2767956
a3	0.1374544
a4	–0.0268917
a5	0.0018094

$$X = \omega = 0.178\theta - 0.277\theta^2 + 0.137\theta^3 - 0.0269\theta^4 + 1.8(10^{-3})\,\theta^5 \quad (3.2)$$

Fatigue failure dimensionless temperature $\boldsymbol{\theta}_f = 6.9$ and $\boldsymbol{\tau}_f = m = 0.45 \rightarrow N_f = (pm)/2\pi = 7.61$ and $T_f = 6.9(60) + 600 = 1014$ [^{0}K] $\approx$ 714 [^{0}C]. Comparing with the temperature rate $\boldsymbol{\tau} = m = [1/100]\,\boldsymbol{\theta}$ (see Chapter 2) the maximum temperature at failure point ($X = \omega = 1$) is $N_f = 1.75 \ll 7.61$ as expected.

Because in a real live situation, the heat flux (and, consequently, the maximum temperature at creep-fatigue failure) is limited ($0 < \theta < \theta$ max) as time tends to infinity, it is interesting to consider, for example, the monotonically increasing exponential function. It should be noted that the inverse function (used in the integral Equation 3.1 of the cyclic creep-fatigue of the composite) in this case is a logarithmic function, i.e., slowly increasing function. The reverse case is also interesting, when the heat flux starts to increase indefinitely at a certain point in time, (an explosion occurs); in this case, the direct and inverse functions are swapped. This is presented below in Examples 3.2 and 3.3.

Example 3.2—Data $\theta = 10(1 - \exp(-a\tau))$; $\tau = m = -(1/a)\ \ln[(10 - \theta)/10]$; $m1 = 1/a[10/(10 - \theta)]$

Differential equations

```
1 d(z11)/d(t) = (1/(1-X)^6)*(exp(t/(1+0.067*t)))*A1*((exp(0.1*m)))*m1*
  (sin (p*m))*z^n
2 d(z12)/d(t) = (1/(1-X)^6)*(exp(t/(1+0.067*t)))*A1*((exp(0.1*m)))*m1*
  (cos (p*m))*z^n
3 d(z21)/d(t) = (1/(1-X)^6)*(exp(t/(1+0.067*t)))*(A2)*(z^s)*((exp
  (0.1*m)))*m1*(sin (p*m))*z^n
4 d(z22)/d(t) = (1/(1-X)^6)*(exp(t/(1+0.067*t)))*(A2)*(z^s)*((exp
  (0.1*m)))*m1*(cos (p*m))*z^n
5 d(z31)/d(t) = (1/(1-X)^2)*(1/k)*(exp(t/(1+0.067*t)))*A3*(z^s)*f3*
  ((exp(0.1*m2)))*m21*((sin (p*m2))^1)*z^n
6 d(z32)/d(t) = (1/(1-X)^2)*(1/k)*(exp(t/(1+0.067*t)))*A3*f3*(z^s)*
  ((exp(0.1*m2)))*m21*1*(cos (p*m2))*z^n
7 d(X1)/d(t) = m1*A*(z^2)/((1-X1)^6)
8 d(X2)/d(t) = m1*A0*(z^2)/((1-X2)^2)
```

Explicit equations

```
1  a = 1
2  m = -(1/a)* ln((10 - t )/ 10)
3  m1 = (1/a)*( 10 / (10 - t))
4  m2 = 0.0868*t - 0.0386*(t^2)+ 0.0080*(t^3) - 0.000756*(t^4) +
   2.6*(10^-5)*(t^5)
5  p = (10^2)
6  fi = 0.05
7  c1 = 3
8  c2 = 5
9  t2 = 1/4
10 t1 = 1/6
11 k = 10
12 n = 1.0
13 s = 1.0
```

14 $E = (0.625-0.375*(\tanh(c1*(t-t1))))*(1-fi)+(k)*(0.625-0.375*(\tanh(c2*(t-t2))))*(fi)$

15 $z1 = (\cos(p*m))*z11-(\sin(p*m))*z12$

16 $z2 = (\cos(p*m))*z21-(\sin(p*m))*z22$

17 $Y1 = (0.15)*(\sin(p*m))*E$

18 $A1 = 1$

19 $deff1 = (1+fi*((t+1)^{0.333}-1))^3 - 1$

20 $d = 1*t$

21 $A2 = 0.1*(\exp(-0.4*t))*((1+2.5*fi+14.7*(fi^2)))*deff1^1$

22 $deff = (fi*d)/(1 + 0.333*(1 - fi)*d)$

23 $z3 = (\cos(p*m2))*z31-(\sin(p*m2))*z32$

24 $X = X1+X2$

25 $A = 2$

26 $A0 = 2$

27 $f3 = $ if $(t<0)$ then (0) else (1)

28 $f31 = (1/16)*t*(8-t)$

29 $f32 = (\exp(-0.04*t))$

30 $f33 = (\sin(3.14*t/4))^2$

31 $f34 = (1/64)*t^2$

32 $f35 = $ if $(t<4)$ then (1) else (0)

33 $f36 = $ if $(t<4)$ then (0) else (1)

34 $z = Y1-(z1+z2)*((\exp(-0.1*m)))-z3*(\exp(-0.1*m2))+0.175*0$

35 $m21 = 0.0868 - 0.0772*t + 0.024*(t^2) - 0.003*(t^3) + 1.3*(10^{-4})*(t^4)$

36 $A3 = deff1$

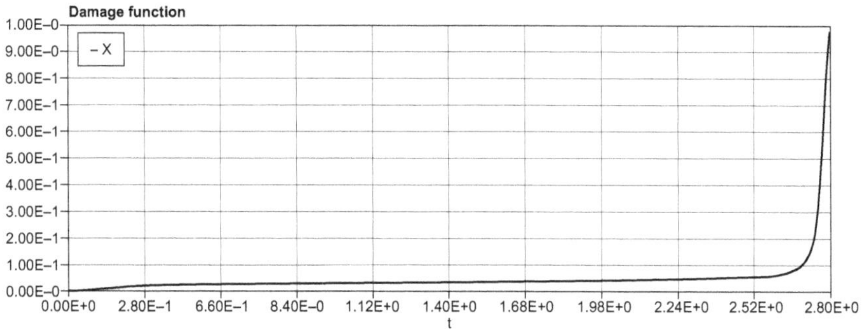

Figure 3.4. Damage function [case $\theta = 10(1 - \exp(-a\tau))$].

Model: X = a1*t + a2*t^2 + a3*t^3 + a4*t^4 + a5*t^5

Variable	Value
a1	0.461752
a2	–1.580522
a3	1.868046
a4	–0.8850077
a5	0.1450593

$$X = \omega = 0.462\theta - 1.58\theta2 + 1.868\theta3 - 0.885\theta4 + 0.145\theta5 \qquad (3.3)$$

Of course, the number of cycles before the failure point in this case strongly depends on the value of the parameter "a". For example, if a = 4 (instead of a = 1), then $\tau_f = m = 0.145 \rightarrow N_f = (pm)/2\pi = 2.308 < 5.157$.

Example 3.3—Inverse Function

Data: See Example 3.2 and $\tau = m = (1 - \exp(-a\theta))$; $m1 = a(\exp(-a\,\theta))$; $0 < m < 1$

Graphically functions 'm' and 'm1'in this case are presented below (compare with previous example—see (Fig. 3.5)).

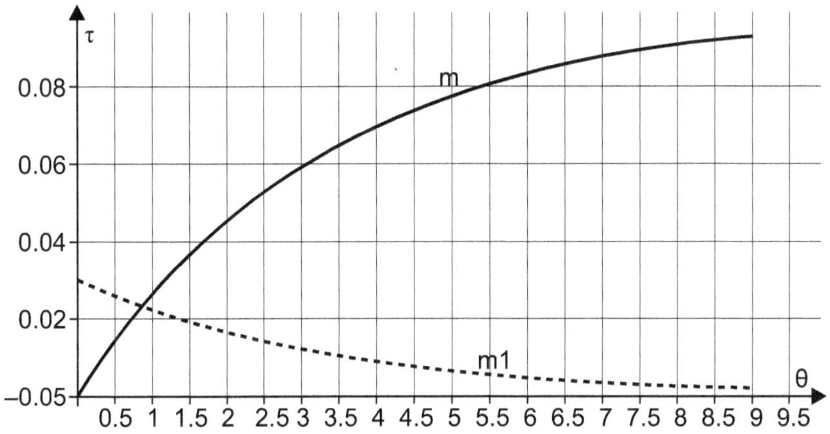

Figure 3.5. Dimensionless functions time-temperature 'm' and first derivative m'.

Differential equations

1 d(z11)/d(t) = (1/(1-X)^6)*(exp(t/(1+0.067*t)))*A1*((exp(0.1*m)))*m1*(sin (p*m))*z^n

2 d(z12)/d(t) = (1/(1-X)^6)*(exp(t/(1+0.067*t)))*A1*((exp(0.1*m)))*m1*(cos (p*m))*z^n

3 d(z21)/d(t) = (1/(1-X)^6)*(exp(t/(1+0.067*t)))*(A2)*(z^s)*((exp (0.1*m)))*m1*(sin (p*m))*z^n

4 d(z22)/d(t) = (1/(1-X)^6)*(exp(t/(1+0.067*t)))*(A2)*(z^s)*((exp (0.1*m)))*m1*(cos (p*m))*z^n

5 d(z31)/d(t) = (1/(1-X)^2)*(1/k)*(exp(t/(1+0.067*t)))*A3*(z^s)*f3*((exp(0.1*m2)))*m21*((sin (p*m2))^1)*z^n

6 d(z32)/d(t) = (1/(1-X)^2)*(1/k)*(exp(t/(1+0.067*t)))*A3*f3*(z^s)*((exp(0.1*m2)))*m21*1*(cos (p*m2))*z^n

7 d(X1)/d(t) = m1*A*(z^2)/((1-X1)^6)

8 d(X2)/d(t) = m1*A0*(z^2)/((1-X2)^2)

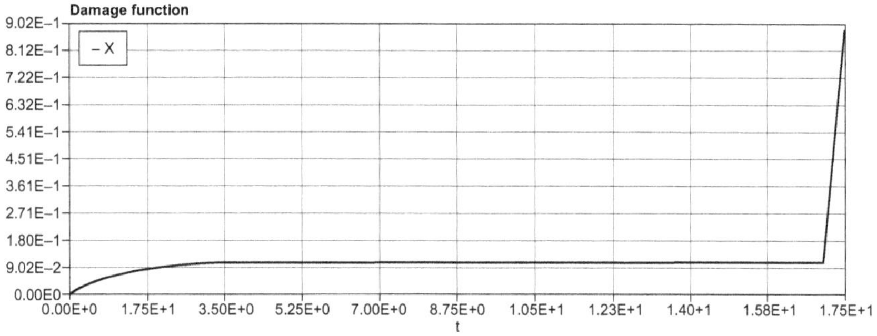

Figure 3.6. Damage function [inverse case to $\theta = 10(1 - \exp(-a\tau))$].

Model: X = a1*t + a2*t^2 + a3*t^3 + a4*t^4 + a5*t^5

Variable	Value
a1	0.1292547
a2	–0.0522579
a3	0.0086778
a4	–0.0006161
a5	1.554E-05

$$X = \omega = 0.129\theta - 0.0522\theta2 + 0.00868\theta3 - 0.000616\theta4 + 1.554(10-5)\,\theta5 \quad (3.4)$$

The maximum fatigue dimensionless temperature $\boldsymbol{\theta}_f = 17.5 >> 6.9$ and $\boldsymbol{\tau f} = m = 1.0 \rightarrow N_f = (pm)/2\pi = 15.9$ and $T_f = 17.5(60) + 600 = 1650$ [^{0}K] $\approx$ 1350 [^{0}C].

Failure occurred at the point where the damage function (see Fig. 3.2) is approaching value X = 1.

As is known, the graph of the function shown above (see Fig. 3.6) is a typical dependence of the heat release during a chain reaction as a result of an explosion.

3.3 Effect of chemical energy on cyclic creep-fatigue process

Consider now another important example that describes the effect of chemical energy on cyclic creep and fatigue behavior of composites and nanocomposites. It has been mentioned in the previous chapter that the third integral in Equation 2.1 is representing the portion of effective stress due to a chemical exothermal reaction (for example, in the case of nanocomposites affected by high temperatures). Therefore, the temperature – time relationship is different from a given external thermal load. In order to obtain this temperature – time function one should establish the "general" mechanism of the chemical reaction (for instance, in "bottom-up "sequence of nanocomposite production), and solve the corresponding kinetic rate equations combined with the energy conservation equation. The examples of such step-by-step procedures can be found by the reader in the author's previous work [8–17]. Below is one example that validates the physical assumptions of Equation 3.1 with respect to the effect of nanomaterials on the overall cyclic creep and fatigue behavior of nanocomposites.

3.3.1 Chemical kinetic effect on nanocomposites creep-fatigue process

Below is an example that is different from the two previous ones: using the law of mass we develop the model for (E)-mediated conversion of complex (S) into a product (P) through the intermediate complex (C). The reaction diagram for that process is.

$$S + E \underset{k_-}{\overset{k_+}{\rightleftharpoons}} C \xrightarrow{k_2} P + E$$

Example 3.4—Each of concentrations S, E, C, and P treated as a dependent variables thar contributes to the total kinetic model. The full set of differential equations in this case is:

$$\frac{d[S]}{d\tau} = k_{-}[C] - k_{+}[S][E]$$
$$\frac{d[E]}{d\tau} = k_{-}[C] - k_{+}[S][E] + k_2[C]$$
$$\frac{d[C]}{d\tau} = k_{+}[S][E] - k_{-}[C] - k_2[C] \quad (3.5)$$
$$\frac{d[P]}{d\tau} = k_2[C]$$
$$\frac{\partial\theta}{\partial\tau} = \delta(1-P)^p P^q \exp(\frac{\theta}{1+\beta\theta}) - \alpha\theta^4$$

d(S)/d(τ) = (C-10*E*S) *4
d(E)/d(τ) = (C-10*E*S+1*C) *4
d(C)/d(τ) = (10*E*S-C-1*C)*4
d(P)/d(τ) = γ*C *4
d(θ)/d(τ)= 20*P^1*(exp(θ)/(1+0.1*θ))))-0.233*(θ)4
S(0)=1; E(0)=1; C(0)=0; P(0)=0; T(0)=0

Differential equations

1 d(S)/d(t) = (C-10*E*S) *4
2 d(E)/d(t) = (C-10*E*S+1*C) *4
3 d(C)/d(t) = (10*E*S-C-1*C)*4
4 d(P)/d(t) = 0.1*C *4
5 d(T)/d(t) = 20*(1-P)^0*P^1*(exp(T/(1+0.1*T)))-0.233*(T^4)

Model: T = a1*t + a2*t^2 + a3*t^3 + a4*t^4

Variable	Value
a1	1.233635
a2	–6.890144
a3	18.32588
a4	–9.825676

θ = 1.234 **τ** – 6.8 **τ** 9 **τ**^2 + 18.32 **τ**^3 – 9.82 6 **τ**^4 (3.6)

The inverse function is:

Model: t = a1*T + a2*T^2 + a3*T^3 + a4*T^4 + a5*T^5

Variable	Value
a1	1.950262
a2	–2.788836
a3	2.050339
a4	–0.6947211
a5	0.0875289

$$\boldsymbol{\tau} = 1.95\ \boldsymbol{\theta} - 2.79\ \boldsymbol{\theta}^2 + 2.05\boldsymbol{\theta}^3 - 0.695\boldsymbol{\theta}^4 + 0.0875\ \boldsymbol{\theta}^5 \tag{3.7}$$

Example 3.5—Data—see Example 3.4 and k+ = 12; k– = 1; $\boldsymbol{\beta} = \mathbf{0.1}$
The reaction diagram for that process given by:

$$A \underset{k_-}{\overset{k_+}{\rightleftarrows}} P$$

For the first order chemical reaction the dimensionless temperature – time relationship described by the system of ODE's is as follows:

d(P)/d(t) = 4*(12*A-1*P)
P(0) = 0
d(A)/d(t) = 4*(–12*A+1*P) (3.8)
A(0) = 1
d(y)/d(t) = 20*P^1*exp(y/(1+.1*y)) – 0.233*y^4
y (0)=0

The coefficient '4' in four Equation (3.8) reflects the scale change of real time independent variable 't' to dimensionless time variable τ that is used in fifth dimensionless ODE describing the conservation of thermal energy of the exothermal unsteady complex autocatalytic chemical reaction. It has been assumed in this example that nanocrystalline materials are grown from h = 1 nm to h = 1.0 mm and the diffusivity coefficient a = 1/4000(1000) = 0.25(10^{-6}) [m2/sec]. Therefore t = [h^2/a]τ = (1.0)2(10-6)/0.25(10-6) = 4τ.

The solution of Equation (3.8) is presented below (using POLYMATH software) via example.

Data: t = [h^2/a] $\boldsymbol{\tau} = \mathbf{4}\ \boldsymbol{\tau}$; ρ = 4000 [kg/m3] α = 10(10^{-6}); c_p = 1000[J/kg*K]; k = 1 [W/m*K]; a = k/c_p ρ [m2/sec]; a = 1/4000(1000) = 0.25(10^{-6});

h = 1.0 mm = 1.0 (10^{-3})m; t = (1)(10^{-6})/0.25(10^{-6}) = 4 τ

k+ =10; k- =1;, and k2 =1 and using computer POLYMATH software we have.

Differential equations

1 d(P)/d(t) = 4*(12*A-1*P)

2 d(A)/d(t) = 4*(-12*A+1*P)

d(y)/d(t) = 20*P^1*exp(y/(1+.1*y)) – 0.233*y^4

Model: t = a1*y + a2*y^2 + a3*y^3 + a4*y^4 + a5*y^5

Variable	Value
a1	0.0868221
a2	–0.0385558
a3	0.0080312
a4	–0.0007564
a5	2.609E-05

$$\tau = m2 = 0.0868\theta - 0.0386\theta^2 + 0.008\theta^3 - 0.000765\theta^4 + 2.61(10^{-5})\,\theta^5 \qquad (3.9)$$

$$\tau' = m21 = 0.0868 - 0.0772\theta 2 + 0.024\theta 2 - 0.00306\theta 3 + 13.05(10^{-5})\,\theta^4 \qquad (3.10)$$

The readers that are interested in more advanced mathematical modeling techniques may read about asymptotic methods; principal components analysis; sensitivity analysis; scaling and the use of non-dimensional variables and parameters, as a means for reducing the number of unknown parameters in a final cyclic creep-fatigue model in references [19 – 21].

3.4 Nanocomposite material under cyclic creep-fatigue conditions

Most kinetics of nanoparticles synthesis processes can be described as a first order chemical reaction except for autocatalytic reactions: they are chemical reactions in which at least one of the products is also a reactant. The rate of these equations for autocatalytic reactions is fundamentally a nonlinear function. This function should include:

Nucleation function (parameters)

It has been found empirically that the logarithm of a reaction rate is proportional to the inverse reaction temperature. This dependency has been described mathematically by the Arrhenius Equation [22,23]:

$$\text{Lnk} = \ln B - Ea/RT \qquad (3.11)$$

where B is called the pre-exponential factor and Ea is the activation energy. Again, the main factors that influence the nucleation rate are the concentration of the reagent and the reaction temperature. There are several basic principles of nucleation: the nucleation rate experiences an exponential increase after a critical energy level has been reached. At high reagent concentrations, the nucleation rate reaches an upper limit. However, it has been found experimentally that the nucleation rate decreases due to increased viscosity [24]. Increasing the temperature lowers the critical value Ea and increases the nucleation rate. These principles can be directly used in the synthesis of nanoparticles.

Cluster developments with rising temperature

The processes of the nucleation and growth of nanoparticles have been described through the La Mer burst nucleation [25,26] and Ostwald ripening [27] to describe the change in the particles size. This process was originally modeled by Reiss [28] with an accepted model being developed by Lifshitz–Slyozov–Wagner, LSW theory [29,30]. The mechanism of nucleation and growth of all nanoparticles is an area of intense interest as it will grant control over the nanoparticles synthesis including particle size and monodispersity. The limits within this field are presented analytically since previously it was the only option to determine any information about the nucleation and growth mechanism. Nucleation is the process whereby nuclei (seeds) act as templates for crystal growth. Primary nucleation is the case of the presence of other crystalline matter as defined by Mullin [31]. This can be used to describe the nucleation of many chemical syntheses [32,33].

Chemical reaction effect

Obviously, for complex reactions, the most common case will be the flow of several simple stages (reactions) with different rates. The difference in rates leads to the fact that the kinetics of product production can be determined by the laws of only one of the stages. For example, for parallel reactions, the rate of the entire process depends on the rate of the fastest stage, and in the sequential ones—the slowest. Accordingly, in the analysis of kinetics in parallel reactions, with a significant difference in the constants, the rate of the slow stage can be neglected, and in successive ones—it is not necessary to determine the fast rate. In successive reactions, the slowest reaction is called the limiting reaction.

To obtain the temperature-time dependence, which is necessary for further analysis of the cyclic creep-fatigue process of the nanocomposites/it is also necessary to add the conservation of energy equation of the entire system (in which the nanocomposites are at a given time), including all external heat sources to the corresponding kinetic equations. For the sake of simplicity of further discussion, such an equation of heat balance will be used (shown in the equation describing the fire dynamics and heat transfer equation that had been used in the author's earlier works [34, 35]).

3.5 Viscosity change effect at high temperature

In order to support the development of new materials required for the design of next generation composites and nanocomposites the high temperature effect of viscosity parameters on cyclical creep-fatigue behavior is required in order to assess the fatigue life and long term durability of advanced polymer matrix composites (PMC's). The viscoelastic constitutive model accounts for some aspects of temperature-dependent viscosity functions and compression loading behavior of composite materials throughout a range of useful temperatures. The research effort detailed in this chapter is an extension of the author's previous work [36] and had two main objectives. The first objective was to explore the effects of elevated temperature on the viscous part of the constitutive model's material parameters. To achieve this goal, the product of two functions has been used (see below): Reynolds function (temperature effect) and Einstein formula (mixture rule: particle size, original form, concentration, special location and more). The second major objective was the verification of the model by comparison of analytical predictions of live fatigue (S – N fatigue curves) and uniaxial test results of composites and nanocomposites. These types of studies should be useful for making comparisons on the effect of temperature between specific laminate types and material systems.

The behavior of nanocomposite materials under cyclic creep-fatigue conditions (similar to Example 3.4) with different nucleation functions (deff and deff1) are presented below:

Example 3.6a—deff1 – nucleation function; m = 0.03θ, m' = 0.03

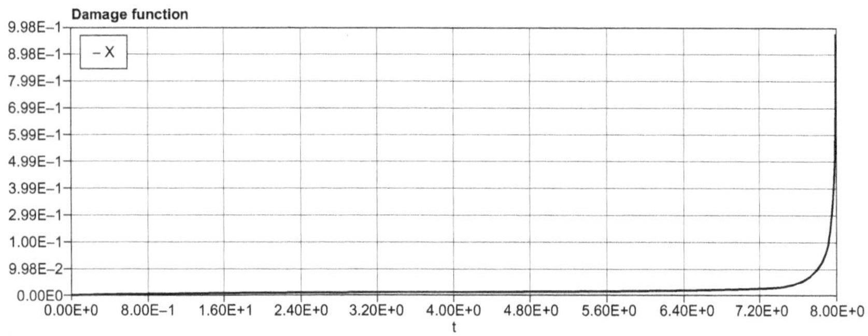

Figure 3.7. Damage function [case deff.1 – nucleation function].

Model: X = a1*t + a2*t^2 + a3*t^3 + a4*t^4 + a5*t^5

Variable	Value
a1	0.1265376
a2	–0.1659588
a3	0.0705627
a4	–0.0118584
a5	0.0006851

$$X = \omega = 0.126\theta - 0.1660\theta 2 + 0.0705\theta 3 - 0.0118\theta 4 + 6.85(10\text{-}4)\ \theta 5 \qquad (3.12)$$

$N_f = 24/2\pi = 3.82$; $T_f = 8(60) + 600 = 1080[^0K] \approx 780\ [^0C]$

Here is another example of the developing process of the nanocomposite material and its effect on cyclic creep-fatigue behavior.

Example 3.6b—(data—see Example 3.6a, but deff instead of deffl); m = 0.03θ, m' = 0.03// The chemical reaction kinetics diagram for the process is: $A \underset{k_-}{\overset{k_+}{\rightleftarrows}} P$.

Differential equations

1 d(z11)/d(t) = (1/(1-X)^6)*(exp(t/(1+0.067*t)))*A1*((exp(0.1*m)))*m1*(sin (p*m))*z^n

2 d(z12)/d(t) = (1/(1-X)^6)*(exp(t/(1+0.067*t)))*A1*((exp(0.1*m)))*m1*(cos (p*m))*z^n

3 d(z21)/d(t) = (1/(1-X)^6)*(exp(t/(1+0.067*t)))*(A2)*(z^s)*((exp (0.1*m)))*m1*(sin (p*m))*z^n

4 d(z22)/d(t) = (1/(1-X)^6)*(exp(t/(1+0.067*t)))*(A2)*(z^s)*((exp (0.1*m)))*m1*(cos (p*m))*z^n

5 d(z31)/d(t) = (1/(1-X)^2)*(1/k)*(exp(t/(1+0.067*t)))*A3*(z^s)*f3*((exp(0.1*m2)))*m21*((sin (p*m2))^1)*z^n

6 d(z32)/d(t) = (1/(1-X)^2)*(1/k)*(exp(t/(1+0.067*t)))*A3*f3*(z^s)*((exp(0.1*m2)))*m21*1*(cos (p*m2))*z^n

7 d(X1)/d(t) = m1*A*(z^2)/((1-X1)^6)

8 d(X2)/d(t) = m1*A0*(z^2)/((1-X2)^2)

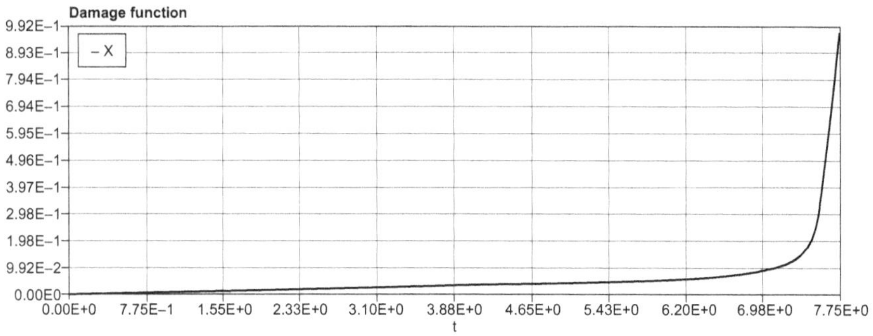

Figure 3.8. Damage function [nucleation function A3 = deff.].

Conclusion: One can see now that $\boldsymbol{\theta}_f = 8.0$ the maximum value of external temperature at failure of nanocomposites and $m = 0.03\theta_f$ are practically the same. The number of cycles for which the accumulated damage from ω1 to ω2 is determined by integrating the kinetic Equation (3.11):

$$N = \int_{\omega_1}^{\omega_2} \frac{d\omega}{f(R, \omega, T)} \tag{3.13}$$

The geometrical interpretation of the damage parameter with $\boldsymbol{\omega} = 1$ corresponds to the case of complete filling of the composite material cross section with micro cracks. In practice, the material becomes unstable and fails when the damage reaches a certain critical value, smaller than one. Due to the significantly nonlinear dependence of the damage parameter on the number of cycles, at the stage preceding failure, the growth rate of $\boldsymbol{\omega}$ increases and rushes to infinity. Therefore, the interval of variation of the damage parameter close to unity, $0.9 < \boldsymbol{\omega} < 1$, corresponds to an insignificant difference in the number of cycles before fatigue failure. This allows determining the STS (or S – N) dependence by integrating relation (3.1) in the range of 0 to the approximately $0.9\theta_{max}$.

The basic experiments to determine the fatigue characteristics of composite materials are experiments on cyclic loading under conditions of a uniaxial stress state. Due to the substantial scatter of experimental data for constructing S – N curves, it is necessary to test a large number of samples at different stress and temperature levels. Theoretical fatigue S – N diagrams corresponding to the found parameters of the damage kinetic equation, are presented in the Fig. 3.2. The experimental data from [9] is also shown for comparison in the same place. After determining the parameters in the proposed model, it can be used to predict the fatigue strength of composites under conditions of a complex stress state. In the calculations and design of parts with variable in volume properties and reinforcement intensity, an important role is acquired by a structured approach, which allows sufficiently flexible control of properties to the location of the composite components by their geometric and physico-mechanical properties (structured parameters). Moreover, from this point of view, the structured approach is also important for understanding the regularities of cyclic deformation of larger structural components, namely monolayer and mono tape, from which the composite product is often formed. This is important not only for direct modeling of fatigue processes, but also for generalizing the results of experimental and computational studies of the behavior of a monolayer under cyclical behavior. On this basis, a possible generalization of experimental data with the construction of phenomenological models of the monolayer, which can then be considered as a structural element in the package of the entire multilayer composite. The results of model computational calculations will help to fill the lack of experimental data when constructing, in particular, phenomenological models. Studying the behavior of a monolayer allows us to generalize a small amount of experimental data. Thus, to build a structured model, the following are necessary:

- Analysis and selection of design models of fatigue strength for different cycle asymmetry factors for nano-materials;
  - collection and systematization of data on changes in the mechanical homogeneous mono-materials, analysis and selection of the most appropriate, sufficiently accurate dependencies describing the fatigue strength during different loading cycles, suitable for use within structured models.
  - analysis of the structured models of the static behavior of a unidirectional composite, with the aim of using individual hypotheses and positions in the development of cyclic deformation models.

All of the above stimulate the development of structured models of fatigue strength, built on hypotheses and assumptions that take into account the specific physical processes in the composite components during cyclic loading, and require, if possible, a minimum amount of experimental data. The use of these models in problems of cyclic loading requires that they have the necessary number of structural parameters for the design, are to a certain extent simple and have sufficient accuracy.

3.6 Analytical expression of crystallization function f3

The fundamental difference between the phenomenological model of cyclic creep-fatigue of nanocomposites and the model of "traditional" composites is the need to take into account four additional factors namely:

1. Allowance for free chemical energy (in the Gibbs equations of thermodynamics), which in turn leads to an additional integral in the creep equation (Equation 2.1 – see Chapter 2).
2. The presence of nanoparticles (dimensions in [nm]) or their conglomerates leads to the need to take into account the dependence of fatigue live on the size of these particles and their evolution (the kinetics of a phase transformation in a material) and its dependence on temperature (time).
3. The mechanical parameters of the nanocomposite structural elements (for example, the effective modulus of elasticity: instantaneous modulus in Equation 2.1; and the hereditary properties of the creep function in Equation 2.1) vary with the temperature field and therefore this should also be reflected in the integral complement to the equation.
4. The mechanical parameters of nanocomposite materials also depend on their chemical composition and the technological method of their production; and hence on the physical kinetics of the chemical reaction itself (at least in a simplified final form). The first factor is taken into account in Equation (3.1) by introducing the activation energy E_a for the nanocomposites and different base temperature T*. The second factor is taken into account by introducing the function f3 (analogy with the well-known HP principle) [38,39] for the dependence of stresses on the size of nanoparticles and the so-called degree of crystallization functions that are encountered in innumerable publications. The best description of the kinetics of phase transformation is given by the so-called Kolmogorov-Johnson-Mehl-Avrami (KJMA) theory. The equation was first derived by Kolmogorov in 1937 and popularized by Melvin Avrami [40–44]. The knowledge of the kinetics of crystallization of nanomaterials is a key point in constructing the cyclic creep-fatigue model that depends on the microstructure. However, from the basic research point of view, it helps to validate the proposed models for phase transformations. The third factor is accounting by introducing the function E - an effective dimensionless instantaneous modulus of elasticity (using the rule of the mixture). Finally, the fourth factor is taken into account by introducing an additional new time-temperature dependence of the chemical reaction based on the chemical kinetics for a given nanomaterial. The process of thermal curing (or degrading) the nanocomposites at each point of the medium is characterized by a monotonically increasing (decreasing)

function f3. Dependence f3 (θ) is a solution of a differential equation of the form:

$$\frac{d(f3)}{d\theta} = f(f3, \theta) \tag{3.14}$$

In general, the analysis of this differential equation is based on the following two main assumptions: (1) The transformation rate at time t during a reaction, df3/dt, can be expressed as a product of two separable functions, one depending solely on the temperature, θ, and the other depending solely on the fraction transformed f3; (2) The temperature function, called the rate constant, follows an Arrhenius type dependency. The relation (3.14) can be chosen in a separable form

$$\frac{d(f3)}{dt} = Ae^{-\frac{E}{RT}} f(f3) \tag{3.15}$$

The Equation (3.15) is the crystallization kinetics equation. In it, θ denotes the dimensionless temperature at the current point of space at the current time. The distribution of temperatures in a continuum medium (see earlier author's work [18]) is found from the solution of the conservation of energy equation

$$\rho \frac{\partial (cT)}{\partial t} = -\operatorname{div}\vec{q} + q \tag{3.16}$$

where ρ is the density, c is the specific heat of the material, and q is the specific heat release power. The heat flux vector q is related to the temperature gradient by the Fourier law

$$q_j = -k_{ij} \frac{\partial T}{\partial x_j} \tag{3.17}$$

On the right-hand side of the Equation (3.17) k_{ij} is the thermal conductivity tensor . Due to the chemical reaction during the curing process, heat generation occurs , which is proportional to the rate of the crystallization of the nanomaterials. For the nanocomposites' analysis the crystallization kinetics equation was taken as the n-th order Prout-Tompkins Equation [15–17] with autocatalysis,

$$\frac{dC}{dt} = Ae^{-\frac{E}{RT}} (1-C)^q C^p \tag{3.18}$$

where A, E, q, p – are parameters, R is the universal gas constant

The heat Equation (3.18) at hand is also separable. This means that the mechanical deformation of the structure does not affect the heat transfer processes occurring in it. Consequently, the temperature field in the computational domain can be found separately, after which the mechanical problem can be solved.

The idea of constructing the cyclic creep-fatigue model of nanocomposites is similar to the creep model of conventional composites (see Chapter 2) with the two major differences of crystallization kinetics function f3 involved in the third integral in Equation (3.1) and the new temperature – time relationship based on the solutions of chemical kinetics equations coupled with the conservation of energy and mass equation [18]. Due to this fact the cyclic creep-fatigue model of nanocomposites is based on the use of different types of functions f3 (see Example 3.7a and Example 3.7b). An analysis of the effect of different types of crystallization (or recrystallization) functions f3 on the fatigue live, N_f is presented below. However, it must be pointed out again that this approach is phenomenological (because of the absence of specific parameters characterizing the function, f3. A similar reasoning also applies to the new temperature – time dependence that should be based on the analysis of the physicochemical kinetics (at least in a very simplified form) of the nanocomposite development. It should be noted that instantaneous stress in Equation 3.1 is assumed for now to be a function of the dimensionless temperature and time and has a cyclic instantaneous stress type, i.e., Y1 = E(θ) [$\sin^2(p\tau)$]. In addition the integral type of cyclic creep-fatigue constitutive equation should be combined with the continuum damage mechanics differential equation. Numerical analysis and results are given below in the form of examples, graphs and tables do not require any significant additional explanations.

Function f3 in this case indicates that chemical free energy is thermally activated from the very beginning (at base temperature T_*) and lasting up to the very end of creep-fatigue process. The effective instantaneous stress and dimensionless glass transitional temperature parameters θ_{gm} and θ_{gf} of the matrix and nanomaterials respectfully are the functions of dimensionless temperature only. The ratio $k = E_f/E_m$ is assumed to be a given constant.

It should be noted that the temperature – time, θ(τ), and the reverse function, τ (θ), are different from m and m1. In the case of a chemical reaction (simple or autocatalytic), they are m2 and m21. Of course, the activation energy and the base temperature are also different, so the dimensionless parameter β could have a different value.

Consider now the effect of different forms of nucleation – crystallization functions f3 on the cyclic creep-fatigue process (see Examples 3.7a – 3.7d).

Example 3.7a—f3 = if (t < 0) then (0) else (1): f3 ≡ 1

1 z1 = (cos(p*m))*z11-(sin(p*m))*z12

2 z2 = (cos(p*m))*z21-(sin(p*m))*z22

3 Y1 = (0.15)*(sin (p*m))*E

Model: X = a1*t + a2*t^2 + a3*t^3 + a4*t^4 + a5*t^5

Variable	Value
a1	0.1189637
a2	–0.1174746
a3	0.0371132
a4	–0.0046434
a5	0.0002001

$$X = \omega = 0.113\theta - 0.118\theta^2 + 0.0371\theta^3 - 0.00464\theta^4 + 2.0(10^{-4})\theta^5 \quad (3.19)$$

Example 3.7b—f3≡ f37: if (t < 0) then (1) else (0)

This case is opposite to that of Example 3.7a: the free chemical energy (third integral in Equation 3.1) does not affect the cyclic creep-fatigue process.

Differential equations

1 d(z11)/d(t) = (1/(1-X)^6)*(exp(t/(1+0.067*t)))*A1*((exp(0.1*m)))*m1 *(sin (p*m))*z^n

2 d(z12)/d(t) = (1/(1-X)^6)*(exp(t/(1+0.067*t)))*A1*((exp(0.1*m)))*m1 *(cos (p*m))*z^n

3 d(z21)/d(t) = (1/(1-X)^6)*(exp(t/(1+0.067*t)))*(A2)*(z^s)* ((exp(0.1*m)))*m1*(sin (p*m))*z^n

4 d(z22)/d(t) = (1/(1-X)^6)*(exp(t/(1+0.067*t)))*(A2)*(z^s)* ((exp(0.1*m)))*m1*(cos (p*m))*z^n

5 d(z31)/d(t) = (1/(1-X)^2)*(1/k)*(exp(t/(1+0.067*t)))*A3*(z^s)*f35* ((exp(0.1*m2)))*m21*((sin (p*m2))^1)*z^n

6 d(z32)/d(t) = (1/(1-X)^2)*(1/k)*(exp(t/(1+0.067*t)))*A3*f35*(z^s)* ((exp(0.1*m2)))*m21*1*(cos (p*m2))*z^n

7 d(X1)/d(t) = m1*A*(z^2)/((1-X1)^6)

8 d(X2)/d(t) = m1*A0*(z^2)/((1-X2)^2)

Model: X = a1*t + a2*t^2 + a3*t^3 + a4*t^4 + a5*t^5

Variable	Value
a1	0.0176571
a2	–0.0163484
a3	0.0050831
a4	–0.0006282
a5	2.674E-05

One can see now that in this case $X_{max} = \omega_{max} = 0.145 << 1$

$$X = \omega = 0.0176\theta + 0.00129\theta^2 - 0.0000942\,\theta^3 + 4.11(10\text{-}6)\theta^4 \tag{3.20}$$

Example 3.7c—is the same as Example 3.7a but: f3 ≡ f32 (similar to the decrease of effective viscosity due to high temperature effect)

Differential equations

1 d(z11)/d(t) = (1/(1-X)^6)*(exp(t/(1+0.067*t)))*A1*((exp(0.1*m)))*m1 *(sin (p*m))*z^n
2 d(z12)/d(t) = (1/(1-X)^6)*(exp(t/(1+0.067*t)))*A1*((exp(0.1*m)))*m1 *(cos (p*m))*z^n
3 d(z21)/d(t) = (1/(1-X)^6)*(exp(t/(1+0.067*t)))*(A2)*(z^s)*((exp (0.1*m)))*m1*(sin (p*m))*z^n
4 d(z22)/d(t) = (1/(1-X)^6)*(exp(t/(1+0.067*t)))*(A2)*(z^s)*((exp (0.1*m)))*m1*(cos (p*m))*z^n
5 d(z31)/d(t) = (1/(1-X)^2)*(1/k)*(exp(t/(1+0.067*t)))*A3*(z^s)*f32* ((exp(0.1*m2)))*m21*((sin (p*m2))^1)*z^n
6 d(z32)/d(t) = (1/(1-X)^2)*(1/k)*(exp(t/(1+0.067*t)))*A3*f32*(z^s)* ((exp(0.1*m2)))*m21*1*(cos (p*m2))*z^n
7 d(X1)/d(t) = m1*A*(z^2)/((1-X1)^6)
8 d(X2)/d(t) = m1*A0*(z^2)/((1-X2)^2)

Damage function [case f3 = f32]

Model: X = a1*t + a2*t^2 + a3*t^3 + a4*t^4 + a5*t^5

Variable	Value
a1	0.1176586
a2	–0.133124
a3	0.060051
a4	–0.0107073
a5	0.0006515

$$X = \omega = 0.1176\theta - 0.133\theta^2 + 0.0060\theta^3 - 0.0107\theta^4 + 6.5150(10^{-5})\theta^5 \quad (3.21)$$

Example 3.7d—is the same as Example 3.7a but: f3 ≡ f31 and fi = φ = 0.5 (The composite is first strengthened under the action of increasing temperature, and then loosened by weakening the bonds between the filler and the matrix.)

Differential equations

1 d(z11)/d(t) = (1/(1-X)^6)*(exp(t/(1+0.067*t)))*A1*((exp(0.1*m)))*m1 *(sin (p*m))*z^n

2 d(z12)/d(t) = (1/(1-X)^6)*(exp(t/(1+0.067*t)))*A1*((exp(0.1*m)))*m1 *(cos (p*m))*z^n

3 d(z21)/d(t) = (1/(1-X)^6)*(exp(t/(1+0.067*t)))*(A2)*(z^s)*((exp (0.1*m)))*m1*(sin (p*m))*z^n

4 d(z22)/d(t) = (1/(1-X)^6)*(exp(t/(1+0.067*t)))*(A2)*(z^s)*((exp (0.1*m)))*m1*(cos (p*m))*z^n

5 d(z31)/d(t) = (1/(1-X)^2)*(1/k)*(exp(t/(1+0.067*t)))*A3*(z^s)*f31* ((exp(0.1*m2)))*m21*((sin (p*m2))^1)*z^n

6 d(z32)/d(t) = (1/(1-X)^2)*(1/k)*(exp(t/(1+0.067*t)))*A3*f31*(z^s)* ((exp(0.1*m2)))*m21*1*(cos (p*m2))*z^n

7 d(X1)/d(t) = m1*A*(z^2)/((1-X1)^6)

8 d(X2)/d(t) = m1*A0*(z^2)/((1-X2)^2)

Damage function [f3 = f31]

Model: X = a1*t + a2*t^2 + a3*t^3 + a4*t^4 + a5*t^5

Variable	Value
a1	0.1157032
a2	–0.1307882
a3	0.0591028
a4	–0.010552
a5	0.0006427

$$X = \omega = 0.116\theta - 0.130\theta^2 + 0.0060\theta^3 - 0.0105\theta^4 + 6.427(10^{-4})\,\theta^5 \quad (3.22)$$

θ_{max} = 8.0 and X = ω = 1.0 and (N_f = 1.27 < 1.75 and θ_{max} = 8.0 < 11.0—see Example 3.7.a above)

3.8 Effect of increase in frequency 'p'

Consider now the effect of an increase in frequency 'p' on the creep-fatigue life of composites and nanocomposites

Example 3.8—is the same as Example 3.7a but: p = 1000 and a_0 = 0.15; a_{st} = 0.175 (The composite is first strengthened under the action of increasing temperature, f31 = 1, and then suddenly loosened.)

Damage function [p = 10^3]

Model: X = a1*t + a2*t^2 + a3*t^3 + a4*t^4 + a5*t^5

Variable	Value
a1	0.1028112
a2	–0.08427
a3	0.0223791
a4	–0.0023535
a5	8.522E-05

$$X = \omega = 0.103\theta - 0.0842\theta^2 + 0.0224\theta^3 - 0.0107\theta^4 + 8.522(10^{-5})\,\theta^5 \qquad (3.23)$$

Θ_{max} = 13.0 and X = ω = 1; (N_f = 20.7 > 1.75—see Example 3.1 above; p = 1000 vs. p = 100)

As expected, with an increase in the frequency p, the lifespan, N_f, of the composite sharply increases (the number of complete oscillation cycles until the failure of a structural element or structure as a whole.

Example 3.8a—same as Example 3.7a but: f3 ≡ f36 and p = 1000 vs. p = 100

Differential equations

1 d(z11)/d(t) = (1/(1-X)^6)*(exp(t/(1+0.067*t)))*A1*((exp(0.1*m)))*m1*(sin (p*m))*z^n
2 d(z12)/d(t) = (1/(1-X)^6)*(exp(t/(1+0.067*t)))*A1*((exp(0.1*m)))*m1*(cos (p*m))*z^n
3 d(z21)/d(t) = (1/(1-X)^6)*(exp(t/(1+0.067*t)))*(A2)*(z^s)*((exp(0.1*m)))*m1*(sin (p*m))*z^n
4 d(z22)/d(t) = (1/(1-X)^6)*(exp(t/(1+0.067*t)))*(A2)*(z^s)*((exp(0.1*m)))*m1*(cos (p*m))*z^n
5 d(z31)/d(t) = (1/(1-X)^2)*(1/k)*(exp(t/(1+0.067*t)))*A3*(z^s)*f36*((exp(0.1*m2)))*m21*((sin (p*m2))^1)*z^n
6 d(z32)/d(t) = (1/(1-X)^2)*(1/k)*(exp(t/(1+0.067*t)))*A3*f36*(z^s)*((exp(0.1*m2)))*m21*1*(cos (p*m2))*z^n
7 d(X1)/d(t) = m1*A*(z^2)/((1-X1)^6)
8 d(X2)/d(t) = m1*A0*(z^2)/((1-X2)^2)

Damage function [f3 =f36 & p = 10^3]

Model: X = a1*t + a2*t^2 + a3*t^3 + a4*t^4 + a5*t^5

Variable	Value
a1	0.1066342
a2	–0.0905321
a3	0.0248762
a4	–0.0027059
a5	0.0001013

$$X = \omega = 0.107\theta - 0.0905\theta 2 + 0.0249\theta 3 - 0.0027\theta 4 + 1.01(10\text{-}4)\theta 5 \quad (3.24)$$

$\boldsymbol{\Theta}_{max}$ = 12.5 T = 12.5(60) +600 = 1350 [^{0}K] ≈ 1050 [0C]; p = 1000 and X = ω = 1.0 (Nf = 19.9 > 1.75—see Example 3.1a above).

Example 3.8—is the same as Example 3.7a but: m = 0.05**θ**; m' = 0.05 and f3 ≡ f36; p = 10000

$$X = \omega = 0.0807\theta - 0.0483\theta^2 + 0.0106\theta^3 - 0.00092\theta^4 + 2.768(10\text{-}5)\,\boldsymbol{\theta}^5 \quad (3.25)$$

$\boldsymbol{\Theta}_{max}$ = 15.5 T_f = 15.5(60) + 600 = 1530[^{0}K] = 1230[^{0}C] and X = ω = 1.0 (Nf = 123.3 > 19.9 —see Example 3.7a above).

Example 3.9—is the same as Example 3.1 but: m = 0.1**θ**; m' = 0.1 and f3 ≡ f36; p = 10000

Differential equations

1 d(z11)/d(t) = (1/(1-X)^6)*(exp(t/(1+0.067*t)))*A1*((exp(0.1*m)))*m1*(sin (p*m))*z^n

2 d(z12)/d(t) = (1/(1-X)^6)*(exp(t/(1+0.067*t)))*A1*((exp(0.1*m)))*m1*(cos (p*m))*z^n

3 d(z21)/d(t) = (1/(1-X)^6)*(exp(t/(1+0.067*t)))*(A2)*(z^s)*((exp(0.1*m)))*m1*(sin (p*m))*z^n

4 d(z22)/d(t) = (1/(1-X)^6)*(exp(t/(1+0.067*t)))*(A2)*(z^s)*((exp(0.1*m)))*m1*(cos (p*m))*z^n

5 d(z31)/d(t) = (1/(1-X)^2)*(1/k)*(exp(t/(1+0.067*t)))*A3*(z^s)*f36*((exp(0.1*m2)))*m21*((sin (p*m2))^1)*z^n

6 d(z32)/d(t) = (1/(1-X)^2)*(1/k)*(exp(t/(1+0.067*t)))*A3*f36*(z^s)*((exp(0.1*m2)))*m21*1*(cos (p*m2))*z^n

7 d(X1)/d(t) = m1*A*(z^2)/((1-X1)^6)

8 d(X2)/d(t) = m1*A0*(z^2)/((1-X2)^2)

Damage function [f3 =f36 & p =10^4, but m = 0.1θ]

Model: X = a1*t + a2*t^2 + a3*t^3 + a4*t^4 + a5*t^5

Variable	Value
a1	0.0895899
a2	–0.0653267
a3	0.020218
a4	–0.0024725
a5	0.0001043

$$X = \omega = 0.0896\theta - 0.0653\theta^2 + 0.0202\theta^3 - 0.00247\theta^4 + 1.043(10\text{-}4)\theta^5 \quad (3.26)$$

Θ_{max} = 11.0 T_f = 11.0(60) + 600 = 1260[°K] = 960[°C] and X = ω = 1.0 (N_f = 278.6 > 19.9—see Example 3.6 above).

As follows from these examples above, an increase in the frequency "p" naturally leads to an increase in the fatigue service life.

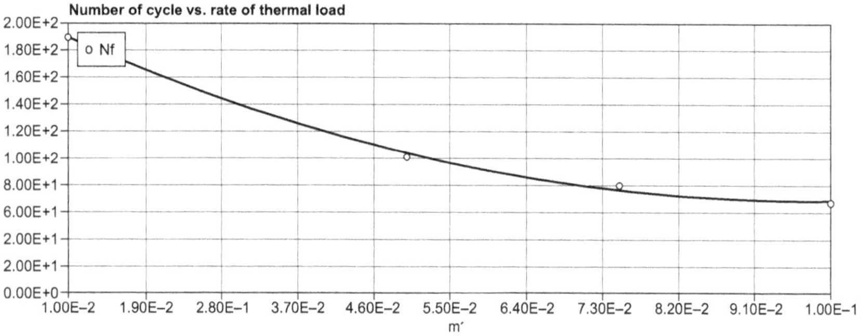

Figure 3.9. Number of Cycles N_f vs. rate of thermal load m' = 0.1 & p = 10^2.

Model: y = a0 + a1*x + a2*x^2

Variable	Value
a0	220.1674
a1	–3147.672
a2	1.628E+04

$$N_f = 220.2 - m'(3147.7) + 16280(m')^2 \quad (3.27)$$

Consider now different cyclic frequencies p and see how the number of cycles to failure changes for different rates of thermal load application.

Model: x = a0 + a1*y + a2*y^2

Variable	Value
a0	0.2550706
a1	–0.0029105
a2	8.522E-06

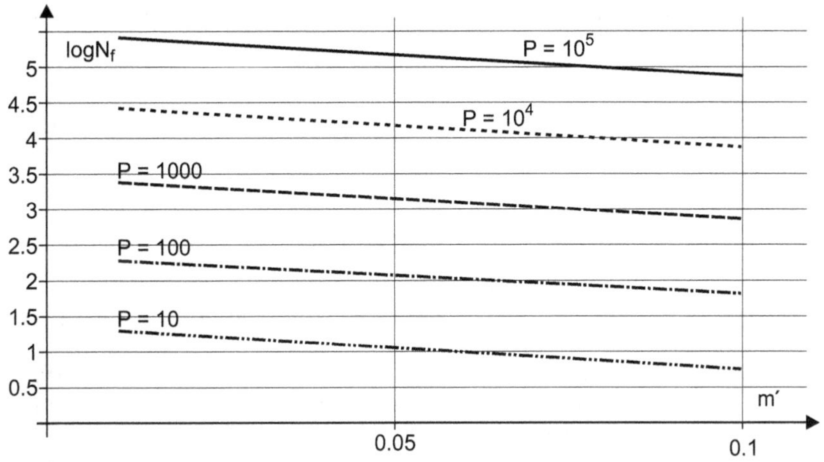

Figure 3.10. Log N_f vs. rate of thermal load m' = 0.1 & p = 10 – 10^5.

References

[1] Moran, M. J. and Shapiro H. N. Fundamentals of Engineering Thermodynamics, Wiley, p. 83, 2008.

[2] Denbigh K. The Principles of Chemical Equilibrium: With Applications in Chemistry and Chemical Engineering. London: Cambridge University Press, 1981.

[3] Razdolsky L. Phenomenological Creep Models of Composites and Nanomaterials, Boca Raton CRC Press, 2019.

[4] Breslavsky D. and Morachkovsky O. New experiments in dynamic creep // Proc. 15th Symp. on Experimental Mechanics of Solids. – Warsaw: Warsaw Techn. Univ., pp. 29–31, 1992.

[5] Sevenois R. D. B. and Van Paepegem W. Fatigue damage modeling techniques for textile composites: Review and comparison with unidirectional composites modeling techniques. ASME Appl. Mech. Rev., 67: 1–12, 2015.

[6] Philippidis T. P. and Passipoularidis V. A. Residual strength after fatigue in composites: Theory vs. experiment. Int. J. Fatigue, 29: 2104–2116, 2007.

[7] Ladeve`ze P. and Lubineau G. A computational meso-damage model for life prediction for laminates. CRC/WP, 3: 432–41, 2003.

[8] Razdolsky L. Phenomenological Creep Models of Composites and Nanomaterials, Boca Raton CRC Press, 2019.

[9] Zhang L., Jin Z., Zhang H., Sui L.H., M.L. and Lu K. 2000. Superheating of confined thin films. Phys. Rev. Lett., 85: 1484—Published.

[10] Hall E. O. The deformation and ageing of mild steel: III discussion of results. Proc. Phys. Soc. London, 64: 747–753, 1951.
[11] Petch N. J. The cleavage strength of polycristals. J. Iron Steel Inst. London, 173: 25, 1953.
[12] Avrami M. Kinetics of Phase Change. I. General Theory. Journal of Chemical Physics., 7(12): 1103–1112, 1939.
[13] Avrami M. Kinetics of Phase Change. II. Transformation-Time Relations for Random Distribution of Nuclei. Journal of Chemical Physics, 8(2): 212–224, 1940.
[14] Avrami M. Kinetics of Phase Change. III. Granulation, Phase Change, and Microstructure. Journal of Chemical Physics., 9(2): 177–184, 1941.
[15] Prout E. G. and Tompkins F. C. The thermal decomposition of potassium permanganate. Trans Faraday Soc., 40: 488–98, 1944.
[16] Prout E. G. and Tompkins F. C. The thermal decomposition of silver permanganate. Trans Faraday Soc., 42: 468–72, 1946.
[17] Jacobs P. W. M. Tompkins FC. Classification and theory of solid reactions. Ch. 7. *In*: Garner WE, editor. Chemistry of the solid state. London: Butterworth Scientific Publication, 1955.
[18] Razdolsky L. Probability Based High Temperature Engineering, Springer Nature Publishing Co. AG Switzerland, 2017.
[19] Laidler K. J. Chemical Kinetics (3rd ed.). Harper & Row. NY., 277 p., 1987.
[20] Nawla H. S. (Ed.). Handbook of Nanostructured Materials and Nanotechnology, Volume 5, Academic Press: New York, NY, USA, 2000.
[21] Boudart M. Kinetics of Chemical Processes. Englewood Cliffs, Prentice-Hall, NJ, USA, 1968.
[22] Fung Y. Foundations of Solid Mechanics (Prentice-Hall, Englewood Cliffs, NJ), 1976.
[23] L. E. Malvern L. Introduction to the Mechanics of a Continuous Medium (Prentice-Hall, Englewood Cliffs, N J), 1969.
[24] Roux J. C., Dekepper P. and Boissonade J. Experimental evidence of nucleation induced transition in a stable chemical system. Phys. Lett. A., 97: 168–170, 1983.
[25] José de Jesús Ibarra-Sánchez, Karla J. Delgado−Carrillo, Ceja-Fdz A., Olivares-Vera D., Sámano A. H. and Cano, M. E. Size Control, Chemical Kinetics, and Theoretical Analysis for the Production of Fe3O4 Nanoparticles with a High Specific Absorption Rate. Industrial & Engineering Chemistry Research, 59(38): 16669–16683, 2020.
[26] Liang Xu Chuanzhen Huang Hanlian Liu Bin Zou Hongtao Zhu Guolong Zhao Jun Wang., Study on the synthesis and growth mechanisms of the refractory ZrC whiskers, International Journal of Refractory Metals and Hard Materials, Volume 42, January (2014), pp. 116–119, 2014.
[27] Bijani S., Schrebler R., Dalchiele E. A., Gabás†L. M. Martínez J. and Ramos-Barrado R. Study of the Nucleation and Growth Mechanisms of Micro- and Nanostructured Cu_2O Thin Films. J. Phys. Chem. C, 115: 43, 2011.
[28] Razdolsky L. Structural Fire Loads: Theory and Principles. McGraw—Hill Co. N.Y., 2012.
[29] Gaurav A. Kamat, Chang Yan, Wojciech T. Osowiecki, Ivan A. Moreno-Hernandez, Marc Ledendecker and Paul A. Alivisatos. Self-Limiting Shell Formation in Cu@Ag Core–Shell Nanocrystals during Galvanic Replacement. The Journal of Physical Chemistry Letters, 11(13), 2020.
[30] Xiaohui Tan, Zhiyu Xue, Hua Zhu, Xin Wang and Dingguo Xu. How to Regulate Nucleation of Biomimetic Hydroxyapatite Nanoparticles on the Surface of Collagen Mimetic Peptides. Molecular Dynamics and Free Energy Investigations. Crystal Growth & Design, 20(7), 2020.
[31] Alexander Pikulin and Nikita Bityurin. Homogeneous Model for the Nanoparticle Growth in Polymer Matrices. The Journal of Physical Chemistry, 2020.

[32] Ting Wang, Wei Lu, Qihua Yang, Sai Li, Xue Yu, Jianbei Qiu, Xuhui Xu and Siu Fung Yu. *In Situ* Observation of Nucleation and Crystallization of a Single Nanoparticle in Transparent Media. The Journal of Physical Chemistry C, 2020.
[33] Jeuss Kirpatric R. Crysral growth from the melt: A review. American Mineralogist, 60: 798–614, 1975.
[34] Razdolsky L. Structural Fire Loads: Theory and Principle. McGaw-Hill, N.Y., 2012.
[35] Razdolsky L. Probability Based Structural Fire Load. Cambridge University Press, London, 2014.
[36] Ritzhaupt-Kleissl E., Haußelt J. and Hanemann T. Thermo-mechanical properties of thermoplastic polymer-nanofiller composites. pp. 87–90. *In*: Proceedings of the 4M 2005 Conference (Multi-MaterialMicro-Manufacture). Elsevier Publisher: Oxford, UK, 2005.
[37] Jordan J., Jacob K. I., Tannenbaum R., Sharaf M.A. and Jasiuk I. Experimental trends in polymer nanocomposites—a review. Mater. Sci. Eng. A, 393: 1–11, 2005.
[38] Hall E. O. The deformation and ageing of mild steel: III discussion of results. Proc. Phys. Soc. London 64: 747–753, 1951.
[39] Petch N. J. The cleavage strength of polycristals. J. Iron Steel Inst. London, 173: 25–28, 1953.
[40] Avrami M. Kinetics of phase change (I). General theory. J. Chem. Phys., 7: 1103–1112, 1939.
[41] Avrami M. Kinetics of phase change (II). Transformation-time relations for random distribution of nuclei. J. Chem. Phys., 8: 212–224, 1940.
[42] Avrami M. Granulation, phase change and microstructure, Kinetics of phase change. III. J. Chem. Phys., 9: 177–184, 1941.
[43] Johnson W. and Mehl K. Reaction kinetics in processes of nucleation and growth. Trans. Am. Inst. Min. Met. Eng., 195: 416–458, 1939.
[44] Kolmogorov A. Static theory of metals crystallization. Izvestia Academia Nauk SSSR. Ser. Mater., 1: 355–359, 1937.

CHAPTER 4

Peculiarities of Phenomenological Models of Nanocomposites

4.1 Introduction

Difficulties in determining the resources of engineering objects are directly related to the complexity of the processes that occur in structural materials under the operating conditions. Understanding the laws of these processes will allow building a reliable mathematical model that will contain specific parameters of the stress-strain state (SSS) that meet the working conditions of the object, and which, ultimately, can become the theoretical basis for creating methods and algorithms for assessing the resources of objects in accordance with individual history of their use. Since the processes of damage accumulation are closely related to the kinetics of the SSS, the accuracy of the calculated estimates of the strength and service life of the structure elements will depend on the degree of certainty with which the ratio determining the mechanics of defective materials (MDM) represent deformation in danger zones of structural elements at predetermined operating conditions. Viscoelastic deformation parameters such as length and type of trajectory, type of stress state, history of its change, and others significantly affect the rate of damage accumulation. Thus, the main goal of research in the field of mechanics of a deformable solid is rather not to clarify the various formulations necessary for determining macroscopic deformations from a given loading history, but to seek to understand the basic laws of phenomena that prepare the ultimate state of the material and structure until it fails.

The life span of a structure operating at elevated temperatures ($T > 0.5T_{mlt}$, T_{mlt} is the melting temperature) and cyclic mechanical stresses is determined mainly by the degradation of the initial strength properties of its materials; low-cycle fatigue and damage accumulation due to creep, which lead to one of the most dangerous types of fracture of structural composite elements—brittle failure [3,4]. The physical processes that cause damage to materials as a result of viscoelastic deformation occur at micro and mesoscale levels and cannot be directly studied by the methods of MDM. The transition from a model that describes events in one of the many micro volumes to a typical engineering representation at the macro level requires the execution of one or another type of averaging. The use of averaging methods based on direct calculation by micro-scale models entails significant complications, which can lead to a significant decrease in the efficiency of numerical calculations if it is necessary to carry out averaging at each time step of integration of the determining relations.

In an alternative version of the approach, generally accepted in the theory of continuum media and implemented in this work, a phenomenological interpretation of models based on macroscopic variables that integrally characterize the structural changes of the material at the micro level is used [1,2]. The peculiarity of the failure of structural elements as a result of low-cycle fatigue (LCF) is the gradual nature of the accumulation of damage from the cyclic action of deformations in areas of structural concentration of damage and areas with a large total margin strength [5,6]. The durability of the composite material is associated with the number and frequency of cycles, as well as with the presence of stress relaxation cycles at elevated constant or changing temperatures T ($0.35T_{mlt} < T < 0.7T_{mlt}$). Moreover, damage accumulation is caused by the presence of developing creep deformations.

Current regulatory methods for assessing the resource of structural elements do not take into account the real processes occurring in the material. The elastic calculation used in the normative approach does not reflect reality either, since the actual characteristics of viscoelastic deformation of the material, largely predetermines the behavior of structural elements with time. Therefore, in the general case of calculating structural strength, time and loading history should be taken into account. Therefore, the defining relations describing the fracture process will affect the strength criterion of the damaged material. In this regard, it becomes necessary to develop new methods for assessing the life of structural elements based on the corresponding equations of thermo- and viscoelasticity, damage accumulation equations and limit state criteria with their comprehensive substantiation through the necessary laboratory and numerical analysis of the processes of deformation and degradation and structural elements under operational conditions.

The defining relations of the mechanics of the damaged medium. The model of the damaged environment consists of three interconnected parts:

- Relation defining the viscoelastic behavior of the material taking into account the influence of the rising temperature process. Analysis of the stress state at variable loads that requires taking into account the anisotropy of the averaged elastic properties of the composite. Micro-structural mechanisms of damage accumulation, including fiber scrap and matrix destruction, fiber matrix cleavage and layering sometimes occur independently.
- At low levels of cyclic loading or in the initial part durability, most types of composites accumulate scattered damage. This damage is spread throughout the tense zone, and gradually reduces the strength and stiffness of the composite. In the later stages of the life cycle, the amount of accumulated damage in some area of the composite can be quite large.
- Due to the variety of structures of composite materials and mechanism of accumulation of damage under cyclic loading it is possible to construct a universal theory of composite fatigue. Although the adhesion behavior of fiber-reinforced composites is significantly different from the behavior of metals, many models have been developed from the known S – N curves.

These models constitute the first class of the so-called 'fatigue-strength'. This approach requires extensive experimental studies and does not take into account the actual mechanisms of damage, such as matrix fractures and fiber breaks [7].

The second class includes phenomenological models for multi-cycle fatigue. These models propose an evolutionary law that describes a gradual degradation of the strength or rigidity of the composite image based on macroscopic properties. Recently, models have been developed based on the concepts of continuum damage mechanics. In models of this type, the damageability is described by some internal parameters of the material. Time-the turn of damage is determined by evolutionary kinetic equations, reflecting the irreversible nature of damage [8].

Continuous damage models introduce scalar, vector, or tensor damage parameters that describe the degradation of every composite material or for structural components. These models are based on physical modeling of the main mechanisms of damage which lead to a macroscopically noticeable degradation of mechanical properties [9].

The main result of all fatigue models is the predicted fatigue life, and each of these three categories uses a different characteristic criterion in order to determine the condition of failure and, as a result, the fatigue life of the composite material considering the continual accumulation model of fatigue damage based on the assumption that the growth rate of the damage parameter depends on the maximum value of the specific energy of elastic deformation of the composite.

4.2 The concept of effective stresses

With a continuous approach to the analysis of the stress state and fatigue of composite products, the material is considered as a homogeneous anisotropic elastic medium [10]. When constructing the model, it is assumed that there are small elastic deformations. The elastic deformation energy functional is quadratic and there is a linear relationship between the stress tensor σ and strain tensor ε. Unlike brittle fracture mechanics, which considers the process of an equilibrium state or the growth of macro cracks, continuous damage hunt uses continuous internal variables that are related to the density of microdefects. The proposed model is based on the concept of effective stress and integrally reflects various types of damage on a micro-scale level, such as the formation and growth of matrix micro-cracks, delamination and other microscopic defects [11].

The complexity and variety of mechanisms for the accumulation of fatigue damage degradation and degradation of the strength properties of the composite make it justified using a scalar internal variable to quantify sanitation of damageability. Damage parameter ω is associated with a decrease in the efficiency area of any cross-section in the vicinity of a given point of the body. By definition, the theoretical value of ω should be within $0 \leq \omega \leq 1$. The effective stress tensor in the case of isotropic damage is introduced as follows.

1. For composites, tensor measures of damage are used. This allows us to take into account the directed, anisotropic nature of the accumulation of fatigue defects. Anisotropic damage models are much more complex than isotropic damage models theoretically, while ensuring compatibility with physical thermodynamic principles of continuum mechanics. To identify the parameters of models of anisotropic damage it is required to carry out a significant number of experiments with complex problems of tests that make it possible to identify the directional character fatigue damage. When considering the geometric interpretation the anisotropic damage tensor is introduced (the second rank damage tensor). By physical considerations, it is symmetrical. Generalization of dependence (4.2) to the case of the damage tensor of the second rank, the symmetric form of the stresses σ at isotropic damageability and the principle of deformation equivalence of the potential are considered.

Helmholtz can be written as:

$$H = U - TS \quad (4.1)$$

where – H is the Helmholtz free energy (SI: joules, CGS: ergs), U is the internal energy of the system (SI: joules, CGS: ergs), T is the absolute temperature (Kelvins), S is the entropy (SI: joules per Kelvin, CGS: ergs per Kelvin). In this case, the thermodynamics of irreversible processes determines the associated damage to the variable ω that shares the energy dissipated during the damage process. Associated damage to the variable ω is determined from the condition that the dispersion is positive.

4.3 Creep-fatigue behavior of nanocomposites

The introduction of a scalar measure of damage determines the choice of mathematics applied.

A technical model for describing the process of accumulation of damage is described. In a complex stress state, the rate of accumulation of damage must depend on the joint invariants of the stress tensors and tensors characterizing the mechanical properties of the composite. This characteristic is taken to be the specific energy of elastic deformation, W_e. Changing the individual components of the stress tensor within the cycle can theoretically occur according to various dependences on time. Striving to reflect cycle characteristics for each component's stress tensor leads to excessive complication of theoretical models. A realistic approach to such situations is to introduce the cycle parameter for the characteristic stress state a_0. As such a parameter is the ratio of the instantaneous amplitude and maximum values of the static stress value. The scalar measure of damageability ω is considered as a function depending on the maximum value of the energy of elastic deformation W_e, the number of load cycles N, the cycle parameter R, the temperature T, material properties and other arguments affecting material fatigue.

Establishing the functional dependence of fatigue strength on the cycle parameter and temperature is a complex problem even for homogeneous materials. When building fatigue models of composites, the most neck spread is the approach in which the material constants are determined from experiments at fixed values of the cycle parameter and temperature, and these characteristics are not included in the damageability equations.

In the proposed model, the hypothesis that the speed of damage accumulation depends on the deformation energy W_e, the ratio of minimum stress to maximum R and the current level of damage is accepted.

The form of the function f, which determines the rate of damage accumulation, should be installed according to the results of experiments on fatigue strength in accordance with the concept of continual mechanics. However, parameter ω can theoretically be controlled with time (the number loading cycles) according to the change in the modulus of elasticity. For practical use of the theory it is more preferable to identify the function's functional dependence based on the results of fatigue strength along the S – N Weller curves [12]. According to the principle of equivalent deformation [13], the specific energy elastic deformation can be expressed in terms of effective stresses. Geometric interpretation of the damage parameter at $\omega = 1$ corresponds to the case of complete filling of the material cross section with micro-cacks. In real practice, the material becomes unstable and collapses, when the damage reaches a certain critical value, less than unity.

Due to the significant nonlinearity of the dependence of the parameter, the number of cycles, at the stage preceding failure, the speed growth increases and rushes to infinity. Therefore, the interval from changes in the damage parameter close to unity, $0.9 < \omega < 1$ correspond to a lower number of cycles. This makes it possible to determine the S - N curve depending on the simplicity of integrating the relation in the range of 0 to 1.

Basic experiments to determine fatigue characteristics of composite materials are experiments on cyclic loading under conditions of the uniaxial stress state. Due to the significance of experimental data to plot S – N curves, it is necessary to test a large number of samples at different levels of stresses.

The capabilities of the proposed model are studied using the example of a four-layer composite with glass fibers and a polyester matrix [14]. Experimental data on fatigue fracture of three types of sample reinforcements, given in [15,16], were processed according to the least squares method for determining the parameters "m" and 'n'. After determining the parameters in the proposed model, it can be used to predict the fatigue strength of composites under conditions of a complex stress state. An energy model is proposed for quantitatively assessing the damageability of composite materials in flat conditions of a stressful state.

The developed model makes it possible to predict the fatigue strength with regard to the influence of the orientation of the principal directions of the stress tensor relative to the planes of elastic symmetry of the material [17,18]. The model uses approaches based on the modern continuum mechanics of damage, and a method for identifying parameters of this model based on the minimum required sets of experimental data [19–21]. The genetic algorithm is a simple, powerful and effective tool used for finding the best solution in a complicated space of design parameters. This algorithm is very different from traditional optimization methods [22]. The genetic algorithm only needs the information based on the objective function, which is its main advantage in comparison with methods based on information gained from objective function derivatives. In contrast to gradient methods, which often fall into a local optimum, this algorithm always finds the global optimum. Thus, the genetic algorithm can serve as an alternative method to classic methods based on mathematical programming. It can be used in many fields where the optimization process is necessary, particularly in the optimal design of a structure made of a fiber-reinforced multilayer composite. Based on the applied theory of probability and the statistical data obtained from the corresponding deterministic solution the theory of fatigue life and resistance is formulated and methods of dimensionless numerical computations for assessing the fatigue resistance are developed. The concept introduces the family of the fatigue curves that are dependent on composite material components data as well as on external and internal forces (all loads are invariant). The work contains numerous results of statistical fatigue examples from various structural materials. For scientists and designers who decide to improve the reliability and durability of parts or the structural systems of unique equipment, fatigue tests are unrealistic. The characteristic criterion of failure in the composite depend on the parameter of the type of stress state and a complete valid analysis for different composites under different loading conditions. Quoting from Talreja at the Second International Conference on Fatigue of Composites (June 2000) [23–25], A reliable and cost-effective fatigue life prediction methodology for composite structures requires a physically based modeling of fatigue damage evolution. An undesirable alternative is an empirical approach. “A major obstacle to developing mechanistic models for composites is the complexity of the fatigue damage mechanisms, both in their geometry and the details of the evolution process. Overcoming this obstacle requires insightful simplification that allows the use of well-developed mechanics’ modeling tools without compromising the essential physical nature of the fatigue process”. From the position of the mechanics of damaged medium (MDM), a mathematical model has been developed that takes into account the processes of damage accumulation in structural elements with a degradation mechanism combining fatigue and creep of the material. An algorithm has been developed for summing damage during the interaction of low-cycle and high-cycle fatigue and creep. Damage accumulation processes are multi-scale and evolutionarily multi-stage. They develop simultaneously at different scale levels: atomic, dislocation, sub-structural, and structural, which means there is a need to combine microscopic, mesoscopic, and macroscopic models [26,27]. For a considerable time, changes occur secretly, in addition, hazardous areas that determine the resource of an element are usually inaccessible to non-destructive testing means. In order to guarantee safe operation and reasonable extension of their service beyond standard terms, it is necessary to control the rate of damage development in the most problematic areas of structural elements, as well as to predict the development of ongoing changes to limit states (calculate the residual life).

Current regulatory methods for assessing the resources of structural elements do not take into account the real processes occurring in the material. The elastic calculation used in the normative approach does not reflect reality either, since the actual characteristics of the viscoelastic deformation of the material largely predetermines the behavior of structural elements with time. Therefore, in the general case of calculating structural strength, time and loading history should be taken into account. Therefore, the defining relations describing the fracture process will affect the strength criterion of the damaged material. In this regard, it becomes necessary to develop new methods for assessing the life of structural elements based on the corresponding equations of thermo- and viscoelasticity, damage accumulation equations and limit state criteria with their comprehensive substantiation through the necessary laboratory and numerical analysis of the processes of deformation and degradation and structural elements under operational conditions.

4.4 Defining the damaged medium mechanics

The model of the damaged environment consists of three interconnected parts:

- relation defining the viscoelastic behavior of the material taking into account the influence of the rising temperature process;
- equation describing the kinetics of damage accumulation;
- criterion strength for the damaged material.

The constitutive relations of thermo-viscoelasticity are based on the following main points:

- The small strain tensor $\boldsymbol{\varepsilon}_{ij}$ is defined in the orthogonal Cartesian coordinate system with local basic vectors, i.e, and the corresponding strain rate tensor $\boldsymbol{\varepsilon}_{ij}$ is the sum of the "instantaneous" and "temporary" components.
- The "instantaneous" component includes elastic strains $\boldsymbol{\varepsilon}_{ij}$ (strain rates $\boldsymbol{\varepsilon}_{ij}$') that are independent of the loading history and are determined by the cyclic state of the process with a given amplitude that are not related to the history of the loading process.
- On the increments of the viscoelastic deformations of structural elements (in contrast to creep deformations), the temperature-time dependence and the history of changes in cyclic external loads have a significant effect.

If the magnitude of stresses, temperature and loading rate are such that the creep effects are significant, the parameters of the material deformation process should be determined taking into account the creep process at the loading stage Δt. To establish the relationship between the creep strain rate tensor $\boldsymbol{\varepsilon}_{ij}$' stress deviator $\boldsymbol{\sigma}_{ij}$', it is assumed that the integral nonlinear cyclic creep-fatigue equation for a uniaxial stress state has the form:

$$E[\theta][a_0\sin(pm)] = \sigma(\theta) + \int_0^\theta A_1 e^{\frac{\theta}{1+\beta\theta}} \frac{1}{(1-\omega)^r} K_1(\theta,\tau)\sigma^n(\tau)m\mathbf{1}d\tau +$$

$$+\int_0^\theta A_2 e^{\frac{\theta}{1+\beta\theta}} \frac{1}{(1-\omega)^r} f_2[\sigma(\tau)]K_2(\theta,\tau)\sigma^n(\tau)m\mathbf{1}d\tau +$$

$$+\int_0^\theta \frac{E_0}{E_1} A_3 e^{\frac{\theta}{1+\beta\theta}} \frac{1}{(1-\omega)^r} f_3[d(\tau)]K_3(\theta,\tau)\sigma(\tau)m\mathbf{21}d\tau + \sigma_{st}$$

$$\frac{d\omega}{d\theta} = A\frac{m\mathbf{1}\sigma^q}{(\mathbf{1}-\omega)^r} + A_0\frac{m\mathbf{1}\sigma^q}{(\mathbf{1}-\omega)^{r\mathbf{1}}};\ \ \omega(0)=\mathbf{0};\ \omega(\theta_f)=\mathbf{1};\ \sigma(\mathbf{0})=\mathbf{0}$$

$$K_1(\theta,\tau) = \sum_{i=1}^N \sin[pm(\theta)-pm(\tau))]\exp(-\alpha_i m\theta)\exp(\alpha_i m\tau)$$

$$K_2(\theta,\tau) = \sum_{i=1}^N \sin[pm(\theta)-pm(\tau)]\exp(-\beta_{i2}m\theta)\exp(\beta_{i2}m\tau)$$

$$K_3(\theta,\tau) = \sum_{i=1}^N \sin[pm\mathbf{2}(\theta)-pm\mathbf{2}(\tau)]\exp(-\beta_{i3}\theta)\exp(\beta_{i3}m\mathbf{2}) \qquad (4.2)$$

$$E = \left(a - b\left(\tanh\left(c\left(\theta-\theta_{gm}\right)\right)\right)\right)(\mathbf{1}-\varphi) + \left(a - b\left(\tanh\left(c\mathbf{1}\left(\theta-\theta_{gf}\right)\right)\right)\right)\varphi / k$$

$f_2(\sigma) = \sigma^s;\ \ s = \mathbf{1},\mathbf{2},\mathbf{3}...,N;$

$k = E_0 / E_1;\ A_1 = \text{ç} = \mathbf{exp}(-\mathbf{b_1}(\beta\theta))$ – Reynolds formula

$A_2 = A_1(\exp(-b_1\beta\theta))((1+2.5\varphi+14.7(\varphi^2))d_{eff1}$

$d=a_1\theta;\ A_3= [f_3(\theta)][d_{eff1}(\theta)];\ d_{eff1}=(1+fi((10t+1)^{1/3}-1))^3 - 1$

$m= 0.01t;\ m1= 0.01$

$m2 = 0.0868\theta - 0.0386(\theta^2)+ 0.0080(\theta^3) - 0.000756(\theta^4) + 2.6(10^{-5})(\theta^5)$

$m21 = 0.0868 - 0.0772\theta+ 0.024(\theta^2) - 0.003(\theta^3) + 1.3(10^{-4})(\theta^4)$

$\beta = \frac{RT_*}{E_a};\ T = \beta T_*\theta + T_*;\ T_*$ – Base temperature $[^0K]$

Consider the cyclic fatigue creep behavior of composites and nanocomposites under high temperatures, i.e., build a S – N fatigue curve by using an analytical method, provided that the operating temperature – time dependence is specified in one of the four ways described in the previous chapter: linearly increasing; non-linearly increasing; exponential and logarithmic function. The main prerequisites and methods for constructing fatigue curves were outlined above. It is necessary to emphasize once again that the proposed model is phenomenological, i.e., theoretically suitable for *any* particular composite or nanocomposite, and therefore all the physical and physicochemical parameters are predefined; and a summary of the calculation and analysis methodology is given below as examples and should be adjusted for each specific case. Concepts used in Equation (4.2) above have been expanded to include the time dependent damage effect on material behavior because of environmental, fabrication, and static load effects.

Example 4.1—Data: $\boldsymbol{\theta} = a\boldsymbol{\tau}$; $\boldsymbol{\tau} = m = [1/a]\,\boldsymbol{\theta}$ and $\boldsymbol{\tau}' = m1 = [1/a]$; $a_0 = 0.15$

Calculated values of DEQ variables

	Variable	Initial value	Minimal value	Maximal value	Final value
1	A	2.	2.	2.	2.
2	A0	2.	2.	2.	2.
3	A1	1.	1.	1.	1.
4	A2	0	0	0.0035427	0.0021916
5	A3	0	0	0.1218714	0.1218714
6	c1	3.	3.	3.	3.
7	c2	5.	5.	5.	5.
8	d	0	0	4.664202	4.664202
9	deff	0	0	0.0942065	0.0942065
10	deff1	0	0	0.1218714	0.1218714
11	E	1.449996	0.3629886	1.449996	0.3629886
12	f3	1.	1.	1.	1.
13	f31	0	0	0.9999986	0.9724273
14	f32	1.	0.8298021	1.	0.8298021
15	f33	0	0	0.9997197	0.2467223
16	f34	0	0	0.3399184	0.3399184
17	f35	1.	0	1.	0
18	f36	0	0	1.	1.
19	f37	0	0	0	0
20	fi	0.05	0.05	0.05	0.05
21	k	10.	10.	10.	10.

22	m	0	0	0.4664202	0.4664202
23	m1	0.1	0.1	0.1	0.1
24	m2	0	0	0.0764684	0.0764684
25	m21	0.0868	0.00073	0.0868	0.0059574
26	n	1.	1.	1.	1.
27	p	100.	100.	100.	100.
28	s	1.	1.	1.	1.
29	t	0	0	4.8	4.8
30	t1	2.	2.	2.	2.
31	t2	4.	4.	4.	4.
32	X	0	0	0.998109	0.998109
33	X1	0	0	0.8279364	0.8279364
34	X2	0	0	0.1701725	0.1701725
35	Y1	0	–0.2141844	0.2174428	0.0252335
36	z	0.175	–0.0375137	10.51892	10.51892
37	z1	0	–10.70519	0.1189807	–10.70519
38	z11	0	–0.522882	2.172E+08	2.172E+08
39	z12	0	–4.153E+08	0.8089179	–4.153E+08
40	z2	0	–0.1060957	0.0002325	–0.1060957
41	z21	0	–0.0007894	4.057E+06	4.057E+06
42	z22	0	–7.757E+06	0.0015802	–7.757E+06
43	z3	0	–7.885E-05	0	–7.885E-05
44	z31	0	–7.469E-06	0.0027422	0.0027422
45	z32	0	–8.052E-07	0.0006569	0.0006569

Differential equations

1 d(z11)/d(t) = (1/(1-X)^6)*(exp(t/(1+0.067*t)))*A1*((exp(0.1*m)))*m1*(sin (p*m))*z^n

```
2   d(z12)/d(t) = (1/(1-X)^6)*(exp(t/(1+0.067*t)))*A1*((exp(0.1*m)))*m1
    *(cos (p*m))*z^n
3   d(z21)/d(t) = (1/(1-X)^6)*(exp(t/(1+0.067*t)))*(A2)*(z^s)*((exp
    (0.1*m)))*m1*(sin (p*m))*z^n
4   d(z22)/d(t) = (1/(1-X)^6)*(exp(t/(1+0.067*t)))*(A2)*(z^s)*((exp
    (0.1*m)))*m1*(cos (p*m))*z^n
5   d(z31)/d(t) = (1/(1-X)^2)*(1/k)*(exp(t/(1+0.067*t)))*A3*(z^s)*f3*
    ((exp(0.1*m2)))*m21*((sin (p*m2))^1)*z^n
6   d(z32)/d(t) = (1/(1-X)^2)*(1/k)*(exp(t/(1+0.067*t)))*A3*f3*(z^s)*
    ((exp(0.1*m2)))*m21*1*(cos (p*m2))*z^n
7   d(X1)/d(t) = m1*A*(z^2)/((1-X1)^6)
8   d(X2)/d(t) = m1*A0*(z^2)/((1-X2)^2)
```

Explicit equations

```
1   m = 0.1*t
2   m1 = 0.1
3   m2 = 0.0868*t - 0.0386*(t^2)+ 0.0080*(t^3) - 0.000756*(t^4) +
    2.6*(10^-5)*(t^5)
4   p = (10^2)
5   fi = 0.05
6   c1 = 3*(1-exp(-0.4*t))^0
7   c2 = 5*(1-exp(-0.4*t))^0
8   t2 = 4
9   t1 = 2
10  k = 10
11  n = 1.0
12  s = 1.0
13  E = (0.625-0.375*(tanh(c1*(t-t1))))*(1-fi)+(k)*(0.625-0.375*
    (tanh(c2*(t-t2))))*(fi)
14  z1 = (cos(p*m))*z11-(sin(p*m))*z12
15  z2 = (cos(p*m))*z21-(sin(p*m))*z22
16  Y1 = (0.15)*(sin (p*m))*E
17  A1 = 1
18  deff1 = (1+fi*((t+1)^0.333-1))^3 - 1
19  d = 1*t
```

```
20  A2 = 0.1*(exp(-0.4*t))*((1+2.5*fi+14.7*(fi^2)))*deff1
21  deff = (fi*d)/(1 + 0.333*(1 - fi)*d)
22  z3 = (cos(p*m2))*z31-(sin(p*m2))*z32
23  X = X1+X2
24  A = 2
25  A0 = 2
26  f3 = if (t<0) then (0) else (1)
27  f31 = (1/16)*t*(8-t)
28  f32 = (exp(-0.04*t))
29  f33 = (sin(3.14*t/4))^2
30  f34 = (1/64)*t^2
31  f35 = if (t<4) then (1) else (0)
32  f36 = if (t<4) then (0) else (1)
33  f37 = if (t<0) then (1) else (0)
34  z = Y1-(z1+z2)*((exp(-0.1*m)))-z3*(exp(-0.1*m2))+0.175*1
35  m21 = 0.0868 - 0.0772*t + 0.024*(t^2) - 0.003*(t^3) + 1.3*(10^-4)*
    (t^4)
36  A3 = deff1
```

Model: X = a1*t + a2*t^2 + a3*t^3 + a4*t^4 + a5*t^5

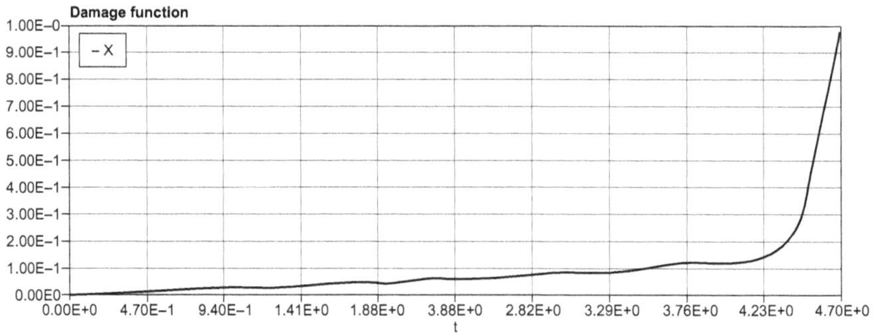

Figure 4.1. Damage Function $a_0 = 0.15$.

Variable	Value
a1	0.2738671
a2	–0.5713533
a3	0.4185196
a4	–0.1215626
a5	0.0121894

$$X = \omega = 0.274\theta - 0.571\theta^2 + 0.418\theta^3 - 0.121\theta^4 + 1.22(10^{-2})\,\theta^5 \quad (4.3)$$

$N_f = 7.16 = m_{max}p/2\pi = 0.45(100)/2\pi = 7.16$ and $T_f = 882[^0K] \approx 582[^0C]$ $(T_f = \beta T_* \theta_{max} + 600)$; $\beta T_* = 0.1\ (600)$; $\theta_{max} = 4.7$

Example 4.2—Data: $\boldsymbol{\theta} = a\boldsymbol{\tau}$; $\boldsymbol{\tau} = m = [1/a]\,\boldsymbol{\theta}$ and $\boldsymbol{\tau}' = m1 = [1/a]$; $a_0 = 0.25$;

Calculated values of DEQ variables

	Variable	Initial value	Minimal value	Maximal value	Final value
1	A	2.	2.	2.	2.
2	A0	2.	2.	2.	2.
3	A1	1.	1.	1.	1.
4	A2	0	0	0.0035426	0.00251
5	A3	0	0	0.1124104	0.1124104
6	c1	3.	3.	3.	3.
7	c2	5.	5.	5.	5.
8	d	0	0	4.123076	4.123076
9	deff	0	0	0.0894635	0.0894635
10	deff1	0	0	0.1124104	0.1124104
11	E	1.449996	0.4472702	1.449996	0.4472702
12	f3	1.	1.	1.	1.
13	f31	0	0	0.9999992	0.9990533
14	f32	1.	0.847959	1.	0.847959
15	f33	0	0	0.9999227	0.0090021

16	f34	0	0	0.2656212	0.2656212
17	f35	1.	0	1.	0
18	f36	0	0	1.	1.
19	f37	0	0	0	0
20	fi	0.05	0.05	0.05	0.05
21	k	10.	10.	10.	10.
22	m	0	0	0.4123076	0.4123076
23	m1	0.1	0.1	0.1	0.1
24	m2	0	0	0.0749249	0.0749249
25	m21	0.0868	0.0007298	0.0868	0.0037877
26	n	1.	1.	1.	1.
27	p	100.	100.	100.	100.
28	s	1.	1.	1.	1.
29	t	0	0	4.2	4.2
30	t1	2.	2.	2.	2.
31	t2	4.	4.	4.	4.
32	X	0	0	0.998239	0.998239
33	X1	0	0	0.8280664	0.8280664
34	X2	0	0	0.1701725	0.1701725
35	Y1	0	–0.3618373	0.3571386	–0.0425176
36	z	0.175	–0.1862692	10.39509	10.39509
37	z1	0	–10.57817	0.2737108	–10.57817
38	z11	0	–1.848E+08	1.290413	–1.848E+08
39	z12	0	–4.494E+08	0.2811756	–4.494E+08
40	z2	0	–0.1163707	0.0004206	–0.1163707
41	z21	0	–3.897E+06	0.0025111	–3.897E+06
42	z22	0	–9.479E+06	0.0004831	–9.479E+06
43	z3	0	–4.164E-05	0	–4.164E-05
44	z31	0	–1.373E-05	0.0009652	0.0009652
45	z32	0	–1.648E-06	0.0004095	0.0004095

Differential equations

1 d(z11)/d(t) = (1/(1-X)^6)*(exp(t/(1+0.067*t)))*A1*((exp(0.1*m)))*m1 *(sin (p*m))*z^n

2 d(z12)/d(t) = (1/(1-X)^6)*(exp(t/(1+0.067*t)))*A1*((exp(0.1*m)))*m1 *(cos (p*m))*z^n

3 d(z21)/d(t) = (1/(1-X)^6)*(exp(t/(1+0.067*t)))*(A2)*(z^s)* ((exp(0.1*m)))*m1*(sin (p*m))*z^n

4 d(z22)/d(t) = (1/(1-X)^6)*(exp(t/(1+0.067*t)))*(A2)*(z^s)* ((exp(0.1*m)))*m1*(cos (p*m))*z^n

5 d(z31)/d(t) = (1/(1-X)^2)*(1/k)*(exp(t/(1+0.067*t)))*A3*(z^s)*f3* ((exp(0.1*m2)))*m21*((sin (p*m2))^1)*z^n

6 d(z32)/d(t) = (1/(1-X)^2)*(1/k)*(exp(t/(1+0.067*t)))*A3*f3*(z^s)* ((exp(0.1*m2)))*m21*1*(cos (p*m2))*z^n

7 d(X1)/d(t) = m1*A*(z^2)/((1-X1)^6)

8 d(X2)/d(t) = m1*A0*(z^2)/((1-X2)^2)

Explicit equations

1 m = 0.1*t

2 m1 = 0.1

3 m2 = 0.0868*t - 0.0386*(t^2)+ 0.0080*(t^3) - 0.000756*(t^4) + 2.6*(10^-5)*(t^5)

4 p = (10^2)

5 fi = 0.05

6 c1 = 3*(1-exp(-0.4*t))^0

7 c2 = 5*(1-exp(-0.4*t))^0

8 t2 = 4

9 t1 = 2

10 k = 10

11 n = 1.0

12 s = 1.0

13 E = (0.625-0.375*(tanh(c1*(t-t1))))*(1-fi)+(k)*(0.625-0.375* (tanh(c2*(t-t2))))*(fi)

14 z1 = (cos(p*m))*z11-(sin(p*m))*z12

15 z2 = (cos(p*m))*z21-(sin(p*m))*z22

```
Y1 = (0.25)*(sin (p*m))*E
A1 = 1
deff1 = (1+fi*((t+1)^0.333-1))^3 - 1
d = 1*t
A2 = 0.1*(exp(-0.4*t))*((1+2.5*fi+14.7*(fi^2)))*deff1
deff = (fi*d)/(1 + 0.333*(1 - fi)*d)
z3 = (cos(p*m2))*z31-(sin(p*m2))*z32
X = X1+X2
A = 2
A0 = 2
f3 = if (t<0) then (0) else (1)
f31 = (1/16)*t*(8-t)
f32 = (exp(-0.04*t))
f33 = (sin(3.14*t/4))^2
f34 = (1/64)*t^2
f35 = if (t<4) then (1) else (0)
f36 = if (t<4) then (0) else (1)
f37 = if (t<0) then (1) else (0)
z = Y1-(z1+z2)*((exp(-0.1*m)))-z3*(exp(-0.1*m2))+0.175*1
m21 = 0.0868 - 0.0772*t + 0.024*(t^2) - 0.003*(t^3) + 1.3*(10^-4)*
(t^4)
A3 = deff1
```

Model: X = a1*t + a2*t^2 + a3*t^3 + a4*t^4 + a5*t^5

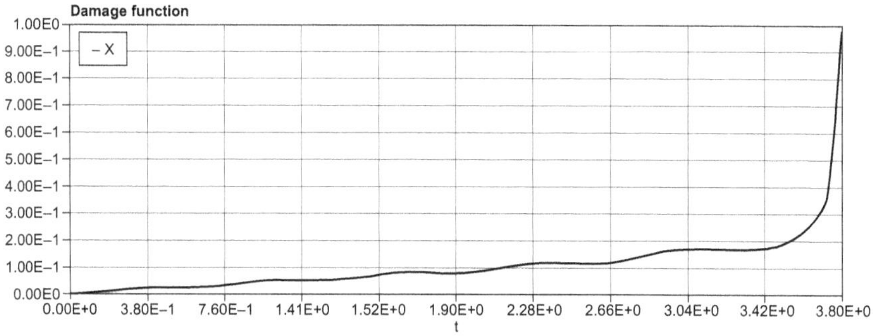

Figure 4.2. Damage Function $a_0 = 0.25$.

Variable	Value
a1	0.3145409
a2	–0.738371
a3	0.6505381
a4	–0.2277467
a5	0.0275667

$$X = \omega = 0.314\theta - 0.738\theta^2 + 0.65\theta^3 - 0.228\theta^4 + 2.76(10^{-2})\,\theta^5 \quad (4.4)$$

$N_f = 6.68 = m_{max}p/2\pi = 0.42(100)/2\pi = 6.68$ and $T_f = 852[^0K] \approx 552[^0C]$ $(T_f = \beta T_* \theta_{max} + 600)$; $\beta T_* = 0.1\ (600)$; $\theta_{max} = 4.2$

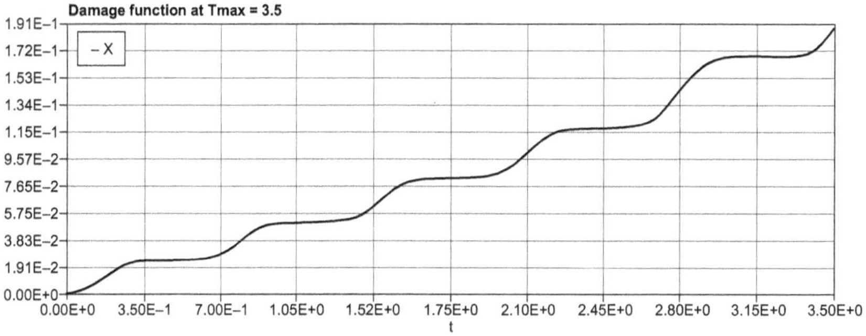

Figure 4.3. Damage Function $\boldsymbol{\theta}_{max} = 3.5$.

Example 4.3—Data: $\boldsymbol{\theta} = a\boldsymbol{\tau}$; $\boldsymbol{\tau} = m = [1/a]\ \boldsymbol{\theta}$ and $\boldsymbol{\tau}' = m1 = [1/a]$; $a_0 = 0.35$

Calculated values of DEQ variables

	Variable	Initial value	Minimal value	Maximal value	Final value
1	A	2.	2.	2.	2.
2	A0	2.	2.	2.	2.
3	A1	1.	1.	1.	1.
4	A2	0	0	0.0035426	0.0030788
5	A3	0	0	0.0940719	0.0940719
6	c1	3.	3.	3.	3.
7	c2	5.	5.	5.	5.
8	d	0	0	3.167164	3.167164

9	deff	0	0	0.0791027	0.0791027
10	deff1	0	0	0.0940719	0.0940719
11	E	1.449996	0.4224744	1.449996	0.4224744
12	f3	1.	1.	1.	1.
13	f31	0	0	0.956649	0.956649
14	f32	1.	0.8810098	1.	0.8810098
15	f33	0	0	0.9999121	0.3714315
16	f34	0	0	0.1567332	0.1567332
17	f35	1.	1.	1.	1.
18	f36	0	0	0	0
19	f37	0	0	0	0
20	fi	0.05	0.05	0.05	0.05
21	k	10.	10.	10.	10.
22	m	0	0	0.3167164	0.3167164
23	m1	0.1	0.1	0.1	0.1
24	m2	0	0	0.0740901	0.0740901
25	m21	0.0868	0.0007289	0.0868	0.000809
26	n	1.	1.	1.	1.
27	p	100.	100.	100.	100.
28	s	1.	1.	1.	1.
29	t	0	0	3.2	3.2
30	t1	2.	2.	2.	2.
31	t2	3.	3.	3.	3.
32	X	0	0	0.9984203	0.9984203
33	X1	0	0	0.8282478	0.8282478
34	X2	0	0	0.1701726	0.1701726
35	Y1	0	–0.5059282	0.5057037	0.0374004
36	z	0.175	–0.3308446	9.89863	9.89863
37	z1	0	–9.869688	0.399293	–9.869688
38	z11	0	–1.210378	1.18E+08	1.18E+08
39	z12	0	–0.8488615	4.514E+08	4.514E+08
40	z2	0	–0.1281774	0.0005591	–0.1281774
41	z21	0	–0.003957	2.901E+06	2.901E+06
42	z22	0	–0.0015077	1.109E+07	1.109E+07

43	z3	0	–5.139E-05	0	–5.139E-05
44	z31	0	–2.4E-05	9.836E-05	9.836E-05
45	z32	0	–3.799E-06	0.0001038	0.0001038

Differential equations

1 d(z11)/d(t) = (1/(1-X)^6)*(exp(t/(1+0.067*t)))*A1*((exp(0.1*m)))*m1*(sin (p*m))*z^n

2 d(z12)/d(t) = (1/(1-X)^6)*(exp(t/(1+0.067*t)))*A1*((exp(0.1*m)))*m1*(cos (p*m))*z^n

3 d(z21)/d(t) = (1/(1-X)^6)*(exp(t/(1+0.067*t)))*(A2)*(z^s)*((exp(0.1*m)))*m1*(sin (p*m))*z^n

4 d(z22)/d(t) = (1/(1-X)^6)*(exp(t/(1+0.067*t)))*(A2)*(z^s)*((exp(0.1*m)))*m1*(cos (p*m))*z^n

5 d(z31)/d(t) = (1/(1-X)^2)*(1/k)*(exp(t/(1+0.067*t)))*A3*(z^s)*f3*((exp(0.1*m2)))*m21*((sin (p*m2))^1)*z^n

6 d(z32)/d(t) = (1/(1-X)^2)*(1/k)*(exp(t/(1+0.067*t)))*A3*f3*(z^s)*((exp(0.1*m2)))*m21*1*(cos (p*m2))*z^n

7 d(X1)/d(t) = m1*A*(z^2)/((1-X1)^6)

8 d(X2)/d(t) = m1*A_0*(z^2)/((1-X2)^2)

Explicit equations

1 m = 0.1*t

2 m1 = 0.1

3 m2 = 0.0868*t - 0.0386*(t^2)+ 0.0080*(t^3) - 0.000756*(t^4) + 2.6*(10^-5)*(t^5)

4 p = (10^2)

5 fi = 0.05

6 c1 = 3*(1-exp(-0.4*t))^0

7 c2 = 5*(1-exp(-0.4*t))^0

8 t2 = 3

9 t1 = 2

10 k = 10

11 n = 1.0

12 s = 1.0

13 E = (0.625-0.375*(tanh(c1*(t-t1))))*(1-fi)+(k)*(0.625-0.375*(tanh(c2*(t-t2))))*(fi)

```
z1 = (cos(p*m))*z11-(sin(p*m))*z12
z2 = (cos(p*m))*z21-(sin(p*m))*z22
Y1 = (0.35)*(sin (p*m))*E
A1 = 1
deff1 = (1+fi*((t+1)^0.333-1))^3 - 1
d = 1*t
A2 = 0.1*(exp(-0.4*t))*((1+2.5*fi+14.7*(fi^2)))*deff1
deff = (fi*d)/(1 + 0.333*(1 - fi)*d)
z3 = (cos(p*m2))*z31-(sin(p*m2))*z32
X = X1+X2
A = 2
A0 = 2
f3 = if (t<0) then (0) else (1)
f31 = (1/16)*t*(8-t)
f32 = (exp(-0.04*t))
f33 = (sin(3.14*t/4))^2
f34 = (1/64)*t^2
f35 = if (t<4) then (1) else (0)
f36 = if (t<4) then (0) else (1)
f37 = if (t<0) then (1) else (0)
z = Y1-(z1+z2)*((exp(-0.1*m)))-z3*(exp(-0.1*m2))+0.175*1
m21 = 0.0868 - 0.0772*t + 0.024*(t^2) - 0.003*(t^3) + 1.3*(10^-4)*
(t^4)
A3 = deff1
```

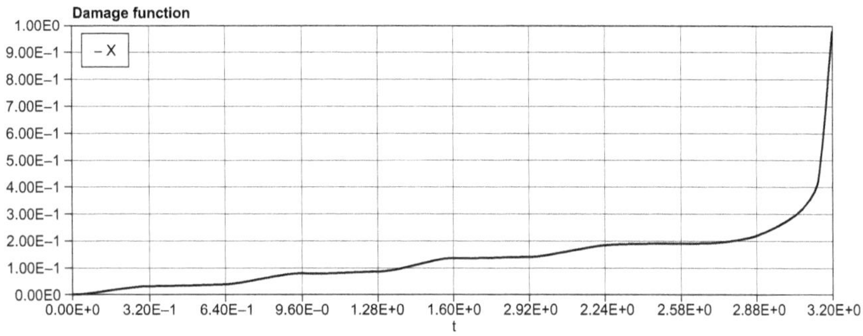

Figure 4.4. Damage Function $a_0 = 0.35$.

Model: X = a1*t + a2*t^2 + a3*t^3 + a4*t^4 + a5*t^5

Variable	Value
a1	0.3995315
a2	–1.053107
a3	1.108208
a4	–0.4632786
a5	0.0667692

$$X = \omega = 0.4\theta - 1.053\theta^2 + 1.108\theta^3 - 0.463\theta^4 + 6.677(10^{-2})\theta^5 \qquad (4.5)$$

$N_f = 5.09 = m_{max}p/2\pi = 0.32(100)/2\pi = 5.09$ and $T_f = 792[^0K] \approx 492[^0C]$ ($T_f = \beta T_* \theta_{max} + 600$); $\beta T_* = 0.1\ (600)$; $\theta_{max} = 3.2$

Example 4.4—Data: $\boldsymbol{\theta} = a\boldsymbol{\tau}$; $\boldsymbol{\tau} = m = [1/a]\ \boldsymbol{\theta}$ and $\boldsymbol{\tau}' = m1 = [1/a]$; $a_0 = 0.45$

Calculated values of DEQ variables

	Variable	Initial value	Minimal value	Maximal value	Final value
1	A	2.	2.	2.	2.
2	A0	2.	2.	2.	2.
3	A1	1.	1.	1.	1.
4	A2	0	0	0.0035427	0.0032874
5	A3	0	0	0.0856854	0.0856854
6	c1	3.	3.	3.	3.
7	c2	5.	5.	5.	5.
8	d	0	0	2.769783	2.769783
9	deff	0	0	0.0738128	0.0738128
10	deff1	0	0	0.0856854	0.0856854
11	E	1.449912	0.3865129	1.449912	0.3865129
12	f3	1.	1.	1.	1.
13	f31	0	0	0.9054103	0.9054103
14	f32	1.	0.8951255	1.	0.8951255
15	f33	0	0	0.9999861	0.6779281
16	f34	0	0	0.1198703	0.1198703
17	f35	1.	1.	1.	1.
18	f36	0	0	0	0

19	f37	0	0	0	0
20	fi	0.05	0.05	0.05	0.05
21	k	10.	10.	10.	10.
22	m	0	0	0.2769783	0.2769783
23	m1	0.1	0.1	0.1	0.1
24	m2	0	0	0.0740252	0.0740252
25	m21	0.0868	0.0009978	0.0868	0.0009978
26	n	1.	1.	1.	1.
27	p	100.	100.	100.	100.
28	s	1.	1.	1.	1.
29	t	0	0	2.8	2.8
30	t1	1.5	1.5	1.5	1.5
31	t2	2.5	2.5	2.5	2.5
32	X	0	0	0.9984361	0.9984361
33	X1	0	0	0.8282636	0.8282636
34	X2	0	0	0.1701726	0.1701726
35	Y1	0	–0.6505544	0.6480101	0.0948094
36	z	0.175	–0.4769554	9.328251	9.328251
37	z1	0	–9.194112	0.3386973	–9.194112
38	z11	0	–0.2303972	2.143E+08	2.143E+08
39	z12	0	–3.296E+08	0.2490518	–3.296E+08
40	z2	0	–0.118662	0.0003833	–0.118662
41	z21	0	–0.000667	5.298E+06	5.298E+06
42	z22	0	–8.149E+06	5.376E-05	–8.149E+06
43	z3	0	–7.226E-05	0	–7.226E-05
44	z31	0	–3.901E-05	8.221E-05	8.221E-05
45	z32	0	–7.524E-06	0.0001202	0.0001202

Differential equations

1 d(z11)/d(t) = (1/(1-X)^6)*(exp(t/(1+0.067*t)))*A1*((exp(0.1*m)))*m1*(sin (p*m))*z^n

2 d(z12)/d(t) = (1/(1-X)^6)*(exp(t/(1+0.067*t)))*A1*((exp(0.1*m)))*m1*(cos (p*m))*z^n

3 d(z21)/d(t) = (1/(1-X)^6)*(exp(t/(1+0.067*t)))*(A2)*(z^s)*((exp(0.1*m)))*m1*(sin (p*m))*z^n

4 d(z22)/d(t) = (1/(1-X)^6)*(exp(t/(1+0.067*t)))*(A2)*(z^s)*((exp(0.1*m)))*m1*(cos (p*m))*z^n

5 d(z31)/d(t) = (1/(1-X)^2)*(1/k)*(exp(t/(1+0.067*t)))*A3*(z^s)*f3*((exp(0.1*m2)))*m21*((sin (p*m2))^1)*z^n

6 d(z32)/d(t) = (1/(1-X)^2)*(1/k)*(exp(t/(1+0.067*t)))*A3*f3*(z^s)*((exp(0.1*m2)))*m21*1*(cos (p*m2))*z^n

7 d(X1)/d(t) = m1*A*(z^2)/((1-X1)^6)

8 d(X2)/d(t) = m1*A0*(z^2)/((1-X2)^2)

Explicit equations

1 m = 0.1*t

2 m1 = 0.1

3 m2 = 0.0868*t - 0.0386*(t^2)+ 0.0080*(t^3) - 0.000756*(t^4) + 2.6*(10^-5)*(t^5)

4 p = (10^2)

5 fi = 0.05

6 c1 = 3*(1-exp(-0.4*t))^0

7 c2 = 5*(1-exp(-0.4*t))^0

8 t2 = 2.5

9 t1 = 1.5

10 k = 10

11 n = 1.0

12 s = 1.0

13 E = (0.625-0.375*(tanh(c1*(t-t1))))*(1-fi)+(k)*(0.625-0.375*(tanh(c2*(t-t2))))*(fi)

14 z1 = (cos(p*m))*z11-(sin(p*m))*z12

15 z2 = (cos(p*m))*z21-(sin(p*m))*z22

16 Y1 = (0.45)*(sin (p*m))*E

17 A1 = 1

18 deffl = (1+fi*((t+1)^0.333-1))^3 - 1

19 d = 1*t

20 A2 = 0.1*(exp(-0.4*t))*((1+2.5*fi+14.7*(fi^2)))*deffl

21 deff = (fi*d)/(1 + 0.333*(1 - fi)*d)

22 z3 = (cos(p*m2))*z31-(sin(p*m2))*z32

23 X = X1+X2

```
A = 2
A0 = 2
f3 = if (t<0) then (0) else (1)
f31 = (1/16)*t*(8-t)
f32 = (exp(-0.04*t))
f33 = (sin(3.14*t/4))^2
f34 = (1/64)*t^2
f35 = if (t<4) then (1) else (0)
f36 = if (t<4) then (0) else (1)
f37 = if (t<0) then (1) else (0)
z = Y1-(z1+z2)*((exp(-0.1*m)))-z3*(exp(-0.1*m2))+0.175*1
m21 = 0.0868 - 0.0772*t + 0.024*(t^2) - 0.003*(t^3) + 1.3*(10^-4)*
(t^4)
A3 = deffl
```

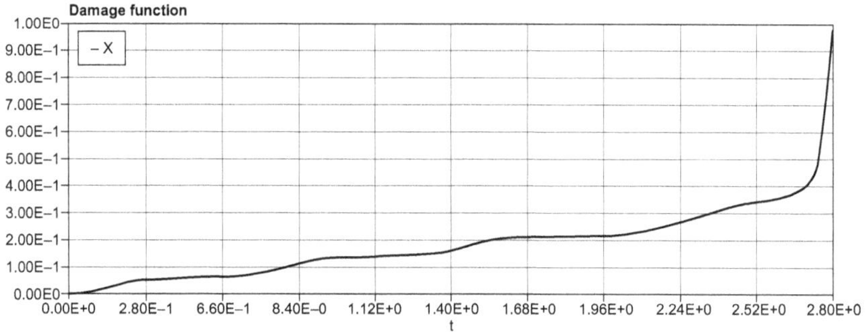

Figure 4.5. Damage Function $a_0 = 0.45$.

Model: X = a1*t + a2*t^2 + a3*t^3 + a4*t^4 + a5*t^5

Variable	Value
a1	0.369885
a2	–0.9139994
a3	1.102724
a4	–0.5343954
a5	0.0896765

$$X = \omega = 0.370 - 0.914\theta^2 + 1.103\theta^3 - 0.534\theta^4 + 8.97(10^{-2})\theta^5 \tag{4.6}$$

$N_f = 4.46 = m_{max}p/2\pi = 0.28(100)/2\pi = 4.46$ and $T_f = 768[^0K] = 468[^0C]$ $(T_f = \beta T_* \theta_{max} + T_*)$ $\beta T_* = 0.1$ (600)); $\theta_{max} = 2.8$

Example 4.5—Data: $\boldsymbol{\theta} = a\boldsymbol{\tau}$; $\boldsymbol{\tau} = m = [1/a]\ \boldsymbol{\theta}$ and $\boldsymbol{\tau}' = m1 = [1/a]$; $a_0 = 0.55$

Calculated values of DEQ variables

	Variable	Initial value	Minimal value	Maximal value	Final value
1	A	2.	2.	2.	2.
2	A0	2.	2.	2.	2.
3	A1	1.	1.	1.	1.
4	A2	0	0	0.0035426	0.003454
5	A3	0	0	0.0765375	0.0765375
6	c1	3.	3.	3.	3.
7	c2	5.	5.	5.	5.
8	d	0	0	2.363973	2.363973
9	deff	0	0	0.0676254	0.0676254
10	deff1	0	0	0.0765375	0.0765375
11	E	1.449708	0.3650291	1.449708	0.3650291
12	f3	1.	1.	1.	1.
13	f31	0	0	0.8327134	0.8327134
14	f32	1.	0.9097741	1.	0.9097741
15	f33	0	0	0.9999993	0.9209925
16	f34	0	0	0.0873182	0.0873182
17	f35	1.	1.	1.	1.
18	f36	0	0	0	0
19	f37	0	0	0	0
20	fi	0.05	0.05	0.05	0.05
21	k	10.	10.	10.	10.
22	m	0	0	0.2363973	0.2363973
23	m1	0.1	0.1	0.1	0.1
24	m2	0	0	0.0734776	0.0734776
25	m21	0.0868	0.0028498	0.0868	0.0028498
26	n	1.	1.	1.	1.
27	p	100.	100.	100.	100.
28	s	1.	1.	1.	1.

29	t	0	0	2.4	2.4
30	t1	1.3	1.3	1.3	1.3
31	t2	1.8	1.8	1.8	1.8
32	X	0	0	0.998549	0.998549
33	X1	0	0	0.8283764	0.8283764
34	X2	0	0	0.1701726	0.1701726
35	Y1	0	–0.7920492	0.7953834	–0.200159
36	z	0.175	–0.6197581	9.315115	9.315115
37	z1	0	–9.43916	0.4500731	–9.43916
38	z11	0	–4.212E+08	0.4799627	–4.212E+08
39	z12	0	–1.536562	3.283E+07	3.283E+07
40	z2	0	–0.1244401	0.0009536	–0.1244401
41	z21	0	–1.092E+07	0.000685	–1.092E+07
42	z22	0	–0.0036633	8.509E+05	8.509E+05
43	z3	0	–0.0001033	0	–0.0001033
44	z31	0	–6.014E-05	7.739E-05	7.739E-05
45	z32	0	–1.301E-05	0.0001611	0.0001611

Differential equations

1 d(z11)/d(t) = (1/(1-X)^6)*(exp(t/(1+0.067*t)))*A1*((exp(0.1*m)))*m1*(sin (p*m))*z^n

2 d(z12)/d(t) = (1/(1-X)^6)*(exp(t/(1+0.067*t)))*A1*((exp(0.1*m)))*m1*(cos (p*m))*z^n

3 d(z21)/d(t) = (1/(1-X)^6)*(exp(t/(1+0.067*t)))*(A2)*(z^s)*((exp(0.1*m)))*m1*(sin (p*m))*z^n

4 d(z22)/d(t) = (1/(1-X)^6)*(exp(t/(1+0.067*t)))*(A2)*(z^s)*((exp(0.1*m)))*m1*(cos (p*m))*z^n

5 d(z31)/d(t) = (1/(1-X)^2)*(1/k)*(exp(t/(1+0.067*t)))*A3*(z^s)*f3*((exp(0.1*m2)))*m21*((sin (p*m2))^1)*z^n

6 d(z32)/d(t) = (1/(1-X)^2)*(1/k)*(exp(t/(1+0.067*t)))*A3*f3*(z^s)*((exp(0.1*m2)))*m21*1*(cos (p*m2))*z^n

7 d(X1)/d(t) = m1*A*(z^2)/((1-X1)^6)

8 d(X2)/d(t) = m1*A0*(z^2)/((1-X2)^2)

Explicit equations

```
1  m = 0.1*t
2  m1 = 0.1
3  m2 = 0.0868*t - 0.0386*(t^2)+ 0.0080*(t^3) - 0.000756*(t^4) + 2.6*(10^-5)*(t^5)
4  p = (10^2)
5  fi = 0.05
6  c1 = 3*(1-exp(-0.4*t))^0
7  c2 = 5*(1-exp(-0.4*t))^0
8  t2 = 1.8
9  t1 = 1.3
10 k = 10
11 n = 1.0
12 s = 1.0
13 E = (0.625-0.375*(tanh(c1*(t-t1))))*(1-fi)+(k)*(0.625-0.375*(tanh(c2*(t-t2))))*(fi)
14 z1 = (cos(p*m))*z11-(sin(p*m))*z12
15 z2 = (cos(p*m))*z21-(sin(p*m))*z22
16 Y1 = (0.55)*(sin (p*m))*E
17 A1 = 1
18 deff1 = (1+fi*((t+1)^0.333-1))^3 - 1
19 d = 1*t
20 A2 = 0.1*(exp(-0.4*t))*((1+2.5*fi+14.7*(fi^2)))*deff1
21 deff = (fi*d)/(1 + 0.333*(1 - fi)*d)
22 z3 = (cos(p*m2))*z31-(sin(p*m2))*z32
23 X = X1+X2
24 A = 2
25 A0 = 2
26 f3 = if (t<0) then (0) else (1)
27 f31 = (1/16)*t*(8-t)
28 f32 = (exp(-0.04*t))
29 f33 = (sin(3.14*t/4))^2
30 f34 = (1/64)*t^2
31 f35 = if (t<4) then (1) else (0)
```

```
f36 = if (t<4) then (0) else (1)
f37 = if (t<0) then (1) else (0)
z = Y1-(z1+z2)*((exp(-0.1*m)))-z3*(exp(-0.1*m2))+0.175*1
m21 = 0.0868 - 0.0772*t + 0.024*(t^2) - 0.003*(t^3) + 1.3*(10^-4)*
(t^4)
A3 = deff1
```

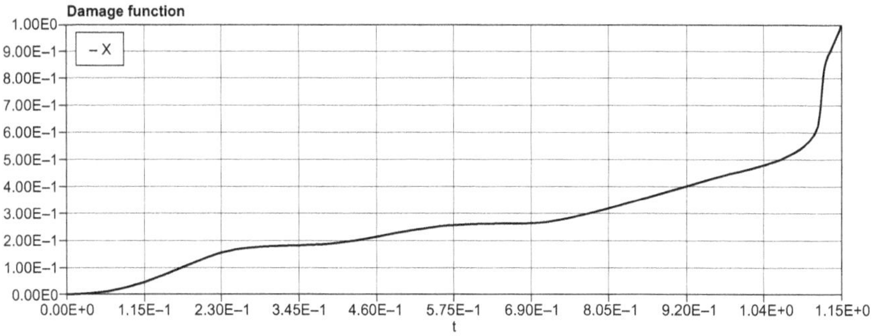

Figure 4.6. Damage Function $a_0 = 0.55$.

Model: X = a1*t + a2*t^2 + a3*t^3 + a4*t^4 + a5*t^5

Variable	Value
a1	0.6130576
a2	–1.832793
a3	2.493532
a4	–1.345676
a5	0.2502034

$$X = \omega = 0.613\theta - 1.833\theta^2 + 2.493\theta^3 - 1.346\theta^4 + 0.25\theta^5 \quad (4.7)$$

$N_f = 3.82 = m_{max}p/2\pi = 0.24(100)/2\pi = 3.82$ and $T_f = 744[^0K] \approx 444[^0C]$ $(T_f = \beta T_* \theta_{max} + T_*)$ $\beta T_* = 0.1\ (600))$; $\theta_{max} = 2.4$

Example 4.6—Data: $\boldsymbol{\theta} = a\boldsymbol{\tau}$; $\boldsymbol{\tau} = m = [1/a]\ \boldsymbol{\theta}$ and $\boldsymbol{\tau}' = m1 = [1/a]$; $a_0 = 0.75$

Calculated values of DEQ variables

	Variable	Initial value	Minimal value	Maximal value	Final value
1	A	2.	2.	2.	2.
2	A0	2.	2.	2.	2.
3	A1	1.	1.	1.	1.
4	A2	0	0	0.0035427	0.0035375
5	A3	0	0	0.0672782	0.0672782
6	c1	3.	3.	3.	3.
7	c2	5.	5.	5.	5.
8	d	0	0	1.981921	1.981921
9	deff	0	0	0.0609079	0.0609079
10	deff1	0	0	0.0672782	0.0672782
11	E	1.416209	0.3656012	1.416209	0.3656012
12	f3	1.	1.	1.	1.
13	f31	0	0	0.7454598	0.7454598
14	f32	1.	0.9237842	1.	0.9237842
15	f33	0	0	0.9997754	0.9997754
16	f34	0	0	0.0613752	0.0613752
17	f35	1.	1.	1.	1.
18	f36	0	0	0	0
19	f37	0	0	0	0
20	fi	0.05	0.05	0.05	0.05
21	k	10.	10.	10.	10.
22	m	0	0	0.1981921	0.1981921
23	m1	0.1	0.1	0.1	0.1
24	m2	0	0	0.0718201	0.0718201
25	m21	0.0868	0.0067187	0.0868	0.0067187
26	n	1.	1.	1.	1.
27	p	100.	100.	100.	100.
28	s	1.	1.	1.	1.
29	t	0	0	2.	2.
30	t1	0.5	0.5	0.5	0.5
31	t2	1.5	1.5	1.5	1.5

32	X	0	0	0.9986085	0.9986085
33	X1	0	0	0.8284359	0.8284359
34	X2	0	0	0.1701726	0.1701726
35	Y1	0	–0.8465433	1.023267	0.2261305
36	z	0.175	–0.6680708	9.019907	9.019907
37	z1	0	–8.673323	0.2560732	–8.673323
38	z11	0	–0.368991	3.297E+08	3.297E+08
39	z12	0	–0.6216207	2.261E+08	2.261E+08
40	z2	0	–0.1178947	0.0001341	–0.1178947
41	z21	0	–0.0008643	8.47E+06	8.47E+06
42	z22	0	–0.0012853	5.809E+06	5.809E+06
43	z3	0	–7.954E-05	0	–7.954E-05
44	z31	0	–4.074E-05	0.0001224	0.0001224
45	z32	0	–1.499E-05	0.000199	0.000199

Differential equations

1 d(z11)/d(t) = (1/(1-X)^6)*(exp(t/(1+0.067*t)))*A1*((exp(0.1*m)))*m1*(sin (p*m))*z^n

2 d(z12)/d(t) = (1/(1-X)^6)*(exp(t/(1+0.067*t)))*A1*((exp(0.1*m)))*m1*(cos (p*m))*z^n

3 d(z21)/d(t) = (1/(1-X)^6)*(exp(t/(1+0.067*t)))*(A2)*(z^s)*((exp(0.1*m)))*m1*(sin (p*m))*z^n

4 d(z22)/d(t) = (1/(1-X)^6)*(exp(t/(1+0.067*t)))*(A2)*(z^s)*((exp(0.1*m)))*m1*(cos (p*m))*z^n

5 d(z31)/d(t) = (1/(1-X)^2)*(1/k)*(exp(t/(1+0.067*t)))*A3*(z^s)*f3*((exp(0.1*m2)))*m21*((sin (p*m2))^1)*z^n

6 d(z32)/d(t) = (1/(1-X)^2)*(1/k)*(exp(t/(1+0.067*t)))*A3*f3*(z^s)*((exp(0.1*m2)))*m21*1*(cos (p*m2))*z^n

7 d(X1)/d(t) = m1*A*(z^2)/((1-X1)^6)

8 d(X2)/d(t) = m1*A0*(z^2)/((1-X2)^2)

Explicit equations

1 m = 0.1*t

2 m1 = 0.1

3 m2 = 0.0868*t - 0.0386*(t^2)+ 0.0080*(t^3) - 0.000756*(t^4) + 2.6*(10^-5)*(t^5)

```
p = (10^2)
fi = 0.05
c1 = 3*(1-exp(-0.4*t))^0
c2 = 5*(1-exp(-0.4*t))^0
t2 = 1.5
t1 = 0.5
k = 10
n = 1.0
s = 1.0
E = (0.625-0.375*(tanh(c1*(t-t1))))*(1-fi)+(k)*(0.625-0.375*
(tanh(c2*(t-t2))))*(fi)
z1 = (cos(p*m))*z11-(sin(p*m))*z12
z2 = (cos(p*m))*z21-(sin(p*m))*z22
Y1 = (0.75)*(sin (p*m))*E
A1 = 1
deff1 = (1+fi*((t+1)^0.333-1))^3 - 1
d = 1*t
A2 = 0.1*(exp(-0.4*t))*((1+2.5*fi+14.7*(fi^2)))*deff1
deff = (fi*d)/(1 + 0.333*(1 - fi)*d)
z3 = (cos(p*m2))*z31-(sin(p*m2))*z32
X = X1+X2
A = 2
A0 = 2
f3 = if (t<0) then (0) else (1)
f31 = (1/16)*t*(8-t)
f32 = (exp(-0.04*t))
f33 = (sin(3.14*t/4))^2
f34 = (1/64)*t^2
f35 = if (t<4) then (1) else (0)
f36 = if (t<4) then (0) else (1)
f37 = if (t<0) then (1) else (0)
z = Y1-(z1+z2)*((exp(-0.1*m)))-z3*(exp(-0.1*m2))+0.175
m21 = 0.0868 - 0.0772*t + 0.024*(t^2) - 0.003*(t^3) + 1.3*(10^-4)*
(t^4)
A3 = deff1
```

Model: X = a1*t + a2*t^2 + a3*t^3 + a4*t^4 + a5*t^5

Variable	Value
a1	0.6703124
a2	–1.751162
a3	2.602095
a4	–1.692005
a5	0.3947546

$$X = \omega = 0.670\theta - 1.751\theta^2 + 2.602\theta^3 - 1.692\theta^4 + 0.395\theta^5 \quad (4.8)$$

$N_f = 3.18 = m_{max}p/2\pi = 0.20(100)/2\pi = 3.18$ and $T_f = 720[^0K] \approx 420[^0C]$ $(T_f = \beta T_* \theta_{max} + T_*)$ $\beta T_* = 0.1\ (600))$; $\theta_{max} = 2.0$

Example 4.6a—Data: $\boldsymbol{\theta} = a\boldsymbol{\tau}$; $\boldsymbol{\tau} = m = [1/a]\ \boldsymbol{\theta}$ and $\boldsymbol{\tau}' = m1 = [1/a]$; $a_0 = 0.95$

Calculated values of DEQ variables

	Variable	Initial value	Minimal value	Maximal value	Final value
1	A	2.	2.	2.	2.
2	A0	2.	2.	2.	2.
3	A1	1.	1.	1.	1.
4	A2	0	0	0.0032179	0.0032179
5	A3	0	0	0.0434788	0.0434788
6	c1	3.	3.	3.	3.
7	c2	5.	5.	5.	5.
8	d	0	0	1.127177	1.127177
9	deff	0	0	0.0415447	0.0415447
10	deff1	0	0	0.0434788	0.0434788
11	E	1.416083	0.3923708	1.416083	0.3923708
12	f3	1.	1.	1.	1.
13	f31	0	0	0.4841807	0.4841807
14	f32	1.	0.9559142	1.	0.9559142
15	f33	0	0	0.598782	0.598782
16	f34	0	0	0.019852	0.019852
17	f35	1.	1.	1.	1.
18	f36	0	0	0	0

19	f37	0	0	0	0
20	fi	0.05	0.05	0.05	0.05
21	k	10.	10.	10.	10.
22	m	0	0	0.1127177	0.1127177
23	m1	0.1	0.1	0.1	0.1
24	m2	0	0	0.0590804	0.0590804
25	m21	0.0868	0.0261881	0.0868	0.0261881
26	n	1.	1.	1.	1.
27	p	100.	100.	100.	100.
28	s	1.	1.	1.	1.
29	t	0	0	1.15	1.15
30	t1	0.5	0.5	0.5	0.5
31	t2	0.8	0.8	0.8	0.8
32	X	0	0	0.9987717	0.9987717
33	X1	0	0	0.8285992	0.8285992
34	X2	0	0	0.1701726	0.1701726
35	Y1	0	–1.061244	1.300347	–0.3586245
36	z	0.175	–0.875677	8.535597	8.535597
37	z1	0	–8.721646	0.1525328	–8.721646
38	z11	0	–3.705E+08	0.4820131	–3.705E+08
39	z12	0	–0.9389388	1.05E+08	1.05E+08
40	z2	0	–0.0963273	0	–0.0963273
41	z21	0	–8.187E+06	0.0006377	–8.187E+06
42	z22	0	–0.0017751	2.321E+06	2.321E+06
43	z3	0	–8.414E-05	0	–8.414E-05
44	z31	0	–0.0001458	2.949E-05	–0.0001458
45	z32	0	–3.24E-05	0.0001405	0.0001405

Differential equations

1 d(z11)/d(t) = (1/(1-X)^6)*(exp(t/(1+0.067*t)))*A1*((exp(0.1*m)))*m1*(sin (p*m))*z^n

2 d(z12)/d(t) = (1/(1-X)^6)*(exp(t/(1+0.067*t)))*A1*((exp(0.1*m)))*m1*(cos (p*m))*z^n

3 d(z21)/d(t) = (1/(1-X)^6)*(exp(t/(1+0.067*t)))*(A2)*(z^s)*((exp(0.1*m)))*m1*(sin (p*m))*z^n

4 d(z22)/d(t) = (1/(1-X)^6)*(exp(t/(1+0.067*t)))*(A2)*(z^s)*((exp(0.1*m)))*m1*(cos (p*m))*z^n

5 d(z31)/d(t) = (1/(1-X)^2)*(1/k)*(exp(t/(1+0.067*t)))*A3*(z^s)*f3*((exp(0.1*m2)))*m21*((sin (p*m2))^1)*z^n

6 d(z32)/d(t) = (1/(1-X)^2)*(1/k)*(exp(t/(1+0.067*t)))*A3*f3*(z^s)*((exp(0.1*m2)))*m21*1*(cos (p*m2))*z^n

7 d(X1)/d(t) = m1*A*(z^2)/((1-X1)^6)

8 d(X2)/d(t) = m1*A0*(z^2)/((1-X2)^2)

Explicit equations

1 m = 0.1*t

2 m1 = 0.1

3 m2 = 0.0868*t - 0.0386*(t^2)+ 0.0080*(t^3) - 0.000756*(t^4) + 2.6*(10^-5)*(t^5)

4 p = (10^2)

5 fi = 0.05

6 c1 = 3*(1-exp(-0.4*t))^0

7 c2 = 5*(1-exp(-0.4*t))^0

8 t2 = 0.8

9 t1 = 0.5

10 k = 10

11 n = 1.0

12 s = 1.0

13 E = (0.625-0.375*(tanh(c1*(t-t1))))*(1-fi)+(k)*(0.625-0.375*(tanh(c2*(t-t2))))*(fi)

14 z1 = (cos(p*m))*z11-(sin(p*m))*z12

15 z2 = (cos(p*m))*z21-(sin(p*m))*z22

16 Y1 = (0.95)*(sin (p*m))*E

17 A1 = 1

18 deff1 = (1+fi*((t+1)^0.333-1))^3 - 1

19 d = 1*t

20 A2 = 0.1*(exp(-0.4*t))*((1+2.5*fi+14.7*(fi^2)))*deff1

21 deff = (fi*d)/(1 + 0.333*(1 - fi)*d)

```
z3 = (cos(p*m2))*z31-(sin(p*m2))*z32
X = X1+X2
A = 2
A0 = 2
f3 = if (t<0) then (0) else (1)
f31 = (1/16)*t*(8-t)
f32 = (exp(-0.04*t))
f33 = (sin(3.14*t/4))^2
f34 = (1/64)*t^2
f35 = if (t<4) then (1) else (0)
f36 = if (t<4) then (0) else (1)
f37 = if (t<0) then (1) else (0)
z = Y1-(z1+z2)*((exp(-0.1*m)))-z3*(exp(-0.1*m2))+0.175*1
m21 = 0.0868 - 0.0772*t + 0.024*(t^2) - 0.003*(t^3) + 1.3*(10^-4)*
(t^4)
A3 = deffl
```

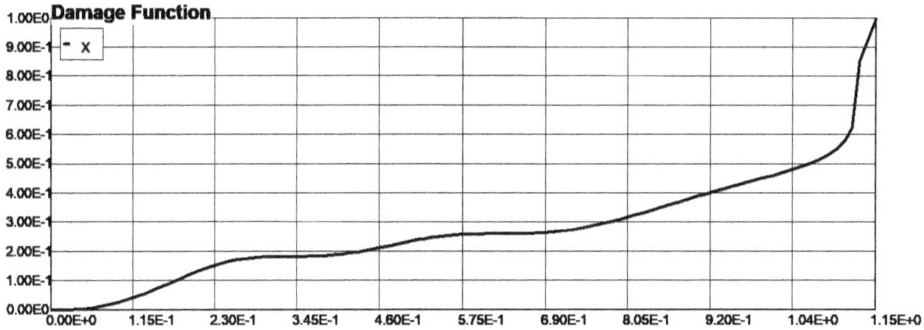

Figure 4.7. Damage Function $a_0 = 0.95$.

Model: X = a1*t + a2*t^2 + a3*t^3 + a4*t^4 + a5*t^5

Variable	Value
a1	0.7714989
a2	–1.799033
a3	4.926278
a4	–6.683018
a5	3.234227

$$X = \omega = 0.67\theta - 1.751\theta^2 + 2.602\theta^3 - 1.692\theta^4 + 0.395\theta^5 \tag{4.9}$$

$N_f = 1.83 = m_{max}p/2\pi = 0.115(100)/2\pi = 1.83$ and $T_f = 669[^0K] \approx 369[^0C]$ $(T_f = \beta T_* \theta_{max} + T_*)$ $\beta T_* = 0.1\ (600))$; $\theta_{max} = 1.15$

Table 4.0. Summary of Examples 4.1 to 4.6 and p =10².

X = ω							
	$\omega = 1.0$ $\theta_f = 4.7$ $N_f = 7.16$	$\omega = 1.0$ $\theta_f = 4.2$ $N_f = 6.67$	$\omega = 1.0$ $\theta_f = 3.2$ $N_f = 5.09$	$\omega = 1.0$ $\theta_f = 2.8$ $N_f = 4.46$	$\omega = 1.0$ $\theta_f = 2.4$ $N_f = 3.82$	$\omega = 1.0$ $\theta_f = 2.0$ $N_f = 3.18$	$\omega = 1.0$ $\theta_f = 1.15$ $N_f = 1.83$

y = a_0; x = N_f

Model: x = a1*y + a2*y^2 + a3*y^3 + a4*y^4 + a5*y^5

Variable	Value
a1	1.186194
a2	–0.446976
a3	0.0419033
a4	0.0026536
a5	–0.0004137

$$a_0 = 1.186N_f - 0.447N_f^2 + 0.0419N_f^3 + 2.65\left(10^{-3}\right)N_f^4 - 4.14\left(10^{-4}\right)N_f^5 \tag{4.10}$$

p = 1; $N_f = 1.875 = m_{max}p/2\pi = 0.1178(100)/2\pi = 1.875$ and $T_f = 672[^0K]$ $(T_f = \beta T_* \theta_{max,f} + 600 = 1.2(60) + 600 = 672[^0K]$; $\beta T_* = 0.1\ (600))$

Table 4.1. Summary of θ_{max} vs. p = 1.

a_0/σ_{st}	0.15	0.25	0.35	0.45	0.55	0.75	0.95
X=ω p=1	$\omega = 1.0$ θ_{max}=5.3 N_f=0.08	$\omega = 1.0$ θ_{max}=5.0 N_f=0.076	$\omega = 1.0$ θ_{max}=4.6 N_f=0.0724	$\omega = 1.0$ θ_{max}=4.5 N_f=0.0688	$\omega = 1.0$ θ_{max}=4.3 N_f=0.065	$\omega = 1.0$ θ_{max}=3.8 N_f=0.0584	$\omega = 1.0$ θ_{max}=3.3 N_f=0.052

$N_f = 1.875 = m_{max}p/2\pi = 0.1178(100)/2\pi = 1.875$ (it can be also counted from Fig. 4.1 or Table 7.1) and $T_f = 672[^0K]$ $(T_f = \beta T_* \theta_{max} + 600 = 1.2(60) + 600 = 672[^0K]$; $\beta T_* = 0.1\ (600))$

Table 4.2. Summary of θ_{max} vs. N_f (p = 10).

a/σ_{st}	0.15	0.25	0.35	0.45	0.55	0.75	0.95
X = ω	$\omega = 1$ $\theta_{max} = 3.0$ $N_f = 0.46$	$\omega = 1$ $\theta_{max} = 2.3$ $N_f = 0.352$	$\omega = 1$ $\theta_{max} = 1.85$ $N_f = 0.286$	$\omega = 1$ $\theta_{max} = 1.65$ $N_f = 0.25$	$\omega = 1$ $\theta_{max} = 1.45$ $N_f = 0.226$	$\omega = 1$ $\theta_{max} = 1.25$ $N_f = 0.192$	$\omega = 1$ $\theta_{max} = 1.1$ $N_f = 0.170$

$N_f = 1.875 = m_{max}p/2\pi = 0.1178(100)/2\pi = 1.875$; and $T_f = 672[^0K]$; $(T_f = \beta T_* \theta_{max} + 600 = 1.2(60) + 600 = 672[^0K]; \beta T_* = 0.1\ (600))$

Table 4.3. Summary of θ_{max} vs. N_f (p = 100).

X = ω	ω = 1.0 $\theta_f = 5.0$ **$N_f = 7.96$**	ω = 1.0 $\theta_f = 4.5$ **$N_f = 7.16$**	ω = 1.0 $\theta_f = 3.8$ **$N_f = 6.05$**	ω = 1.0 $\theta_f = 2.9$ **$N_f = 4.6$**	ω = 1.0 $\theta_f = 2.2$ **$N_f = 3.5$**	ω = 1.0 $\theta_f = 1.5$ **$N_f = 2.39$**	ω = 1.0 $\theta_f = 1.2$ **$N_f = 1.91$**

$N_f = 19.25 = m_{max}p/2\pi = 0.121(1000)/2\pi = 19.25$ and $T_f = 675[^0K]$ $(T_f = \beta T_* \theta_{max} + 600 = 1.25(60) + 600 = 675[^0K]; \beta T_* = 0.1\ (600))$

Table 4.4. Summary of θ_{max} vs. N_f (p = 1000).

a/σ_{st}	0.15	0.25	0.35	0.45	0.55	0.75	0.95
X = ω	ω = 1.0 $\theta_{max} = 6.4$ $N_f = 97.6$	ω = 1.0 $\theta_{max} = 5.5$ $N_f = 84.4$	ω = 1.0 $\theta_{max} = 4.2$ $N_f = 65.9$	ω = 1.0 $\theta_{max} = 3.0$ $N_f = 47.7$	ω = 1.0 $\theta_{max} = 2.2$ $N_f = 35.0$	ω = 1.0 $\theta_{max} = 1.6$ $N_f = 25.5$	ω = 1.0 $\theta_{max} = 1.25$ $N_f = 19.3$

$N_f = 192.5 = m_{max}p/2\pi = 0.121(10000)/2\pi = 192.5$ and $T_f = 675[^0K]$ $(T_f = \beta T_* \theta_{max} + 600 = 1.25(60) + 600 = 675[^0K]; \beta T_* = 0.1\ (600))$

Table 4.5. Summary of θ_{max} vs. N_f (p = 10000).

a/σ_{st}	0.15	0.25	0.35	0.45	0.55	0.75	0.95
X = ω	ω = 1 $\theta_{max} = 6.6$ $N_f = 1001.1$	ω = 1 $\theta_{max} = 5.6$ $N_f = 864.1$	ω = 1 $\theta_{max} = 4.4$ $N_f = 671.6$	ω = 1 $\theta_{max} = 3.1$ $N_f = 477.5$	ω = 1 $\theta_{max} = 2.2$ $N_f = 346.9$	ω = 1 $\theta_{max} = 1.6$ $N_f = 241.9$	ω = 1 $\theta_{max} = 1.25$ $N_f = 192.6$

$N_f = 1925 = m_{max}p/2\pi = 0.121(100000)/2\pi = 1925$ and $T_f = 672[^0K]$ $(T_f = \beta T_* \theta_{max} + 600 = 1.21(60) + 600 = 672[^0K]; \beta T_* = 0.1\ (600))$

Table 4.6. Summary of θ_{max} vs. N_f (p = 100000).

a/σ_{st}	0.15	0.25	0.35	0.45	0.55	0.75	0.95
X = ω	ω = 1 $\theta_{max} = 6.6$ $N_f = 10{,}011$	ω = 1 $\theta_{max} = 5.6$ $N_f = 8{,}641$	ω = 1 $\theta_{max} = 4.4$ $N_f = 6{,}716$	ω = 1 $\theta_{max} = 3.1$ $N_f = 4{,}775$	ω = 1 $\theta_{max} = 2.2$ $N_f = 3{,}469$	ω = 1 $\theta_{max} = 1.6$ $N_f = 2{,}419$	ω = 1 $\theta_{max} = 1.25$ $N_f = 1{,}926$

$N_f = 19250 = m_{max}p/2\pi = 0.121(100000)/2\pi = 19250$; and $T_f = 675[^0K]$ $(T_f = \beta T_* \theta_{max} + 600 = 1.25(60) + 600 = 675[^0K]; \beta T_* = 0.1\ (600))$

Table 4.7. Summary of θ_{max} vs. N_f (p = 1,000,000).

a/σ_{st}	0.15	0.25	0.35	0.45	0.55	0.75	0.95
X = ω	ω = 1 $\theta_{max} = 6.6$ $N_f =$ 100,110	ω = 1 $\theta_{max} = 5.6$ $N_f =$ 86,410	ω = 1 $\theta_{max} = 4.4$ $N_f =$ 67,160	ω = 1 $\theta_{max} = 3.1$ $N_f =$ 47,750	ω = 1 $\theta_{max} = 2.2$ $N_f =$ 34,690	ω = 1 $\theta_{max} = 1.6$ $N_f =$ 24,190	ω = 1 $\theta_{max} = 1.25$ $N_f =$ 19,250

$N_f = 192500 = m_{max}p/2\pi = 0.121(10000000)/2\pi = 192500$ and $T_f = 675[^0K]$ $(T_f = \beta T_* \theta_{max} + 600 = 1.25(60) + 600 = 675[^0K]; \beta T_* = 0.1\ (600))$

Table 4.8. Summary of θ_{max} vs. N_f (p = 10,000,000).

a/σ_{st}	0.15	0.25	0.35	0.45	0.55	0.75	0.95
X = ω	ω = 1 $\theta_{max} = 6.6$ $N_f =$ 1,001,100	ω = 1 $\theta_{max} = 5.6$ $N_f =$ 864,100	ω = 1 $\theta_{max} = 4.4$ $N_f =$ 671,600	ω = 1 $\theta_{max} = 3.1$ $N_f =$ 477,500	ω = 1 $\theta_{max} = 2.2$ $N_f =$ 346,900	ω = 1 $\theta_{max} = 1.6$ $N_f =$ 241,900	ω = 1 $\theta_{max} = 1.25$ $N_f =$ 192,500

Table 4.9. Summary of all θ_{max} vs. N_f (p =1 – 10,000,000).

a/σ_{st}	0.15	0.25	0.35	0.45	0.55	0.75	0.95
X = ω p = 1	ω = 1.0 $m_{max} = 0.50$ $\theta_{max} = 5.3$ $N_f = 0.08$	ω = 1.0 $m_{max} = 0.48$ $\theta_{max} = 5.0$ $N_f = 0.076$	ω = 1.0 $m_{max} = 0.455$ $\theta_{max} = 4.6$ $N_f = 0.0724$	ω = 1.0 $m_{max} = 0.43$ $\theta_{max} = 4.5$ $N_f =$ 0.0688	ω = 1.0 $m_{max} = 0.409$ $\theta_{max} = 4.3$ $N_f = 0.065$	ω = 1.0 $m_{max} =$ 0.367 $\theta_{max} = 3.8$ $N_f = 0.0584$	ω = 1.0 $m_{max} = 0.327$ $\theta_{max} = 3.3$ $N_f = 0.052$
p = 10	ω = 1 $m_{max} = 0.290$ $\theta_{max} = 3.0$ $N_f = 0.46$	ω = 1 $m_{max} =$ 0.221 $\theta_{max} = 2.3$ $N_f = 0.352$	ω = 1 $m_{max} = 0.180$ $\theta_{max} = 1.85$ $N_f = 0.286$	ω = 1 $m_{max} =$ 0.157 $\theta_{max} = 1.65$ $N_f = 0.25$	ω = 1 $m_{max} = 0.142$ $\theta_{max} =1.45$ $N_f =0.226$	ω = 1 $m_{max} =$ 0.121 $\theta_{max} =1.25$ $N_f = 0.192$	ω = 1 $m_{max} = 0.107$ $\theta_{max} = 1.1$ $N_f = 0.170$
X = ω p = 10^2	ω = 1.0 $m_{max} =0.5$ $\theta_{max} =5.0$ $N_f = 7.96$	ω = 1.0 $m_{max} =$ 0.45 $\theta_{max} = 4.5$ $N_f = 7.16$	ω = 1.0 $m_{max} = 0.38$ $\theta_{max} = 3.8$ $N_f = 6.05$	ω = 1.0 $m_{max} = 0.29$ $\theta_{max} = 2.9$ $N_f = 4.6$	ω = 1.0 $m_{max} = 0.22$ $\theta_{max} =2.2$ $N_f =3.5$	ω = 1.0 $m_{max} = 0.15$ $\theta_{max} =1.5$ $N_f =2.39$	ω = 1.0 $m_{max} = 0.12$ $\theta_{max} =1.2$ $N_f = 1.91$
X = ω p = 10^3	ω = 1.0 m=0.61 $\theta_{max} =6.4$ $N_f = 97.6$	ω = 1.0 m= 0.53 $\theta_{max} = 5.5$ $N_f = 84.4$	ω = 1.0 m= 0.414 $\theta_{max} =4.2$ $N_f =65.9$	ω = 1.0 m= 0.3 $\theta_{max} = 3.0$ $N_f = 47.7$	ω = 1.0 m= 0.22 $\theta_{max} = 2.2$ $N_f =35.0$	ω = 1.0 m=0.16 $\theta_{max} = 1.6$ $N_f =25.5$	ω = 1.0 m= 0.12 $\theta_{max} = 1.25$ $N_f = 19.3$
X = ω p = 10^4	ω = 1 $m_{max} = 0.629$ $\theta_{max} = 6.6$ $N_f =1001.1$	ω = 1 $m_{max} = 0.54$ $\theta_{max} = 5.6$ $N_f = 864.1$	ω = 1 $m_{max} = 0.422$ $\theta_{max} = 4.4$ $N_f =671.6$	ω = 1 $m_{max} = 0.3$ $\theta_{max} = 3.1$ $N_f = 477.5$	ω = 1 $m_{max} = 0.218$ $\theta_{max} = 2.2$ $N_f = 346.9$	ω = 1 $m_{max} =0.152$ $\theta_{max} = 1.6$ $N_f = 241.9$	ω = 1 $m_{max} = 0.121$ $\theta_{max} = 1.25$ $N_f = 192.6$
X = ω p = 10^5	ω = 1 $m_{max} = 0.629$ $\theta_{max} = 6.6$ $N_f = 10,011$	ω = 1 $m_{max} = 0.54$ $\theta_{max} = 5.6$ $N_f = 8,641$	ω = 1 $m_{max} = 0.422$ $\theta_{max} = 4.4$ $N_f =6,716$	ω = 1 $m_{max} = 0.3$ $\theta_{max} = 3.1$ $N_f = 4,775$	ω = 1 $m_{max} = 0.218$ $\theta_{max} = 2.2$ $N_f =3,469$	ω = 1 $m_{max} =$ 0.152 $\theta_{max} = 1.6$ $N_f = 2,419$	ω = 1 $m_{max} = 0.121$ $\theta_{max} = 1.25$ $N_f = 1,926$
X = ω p = 10^6	ω = 1 $m_{max} = 0.629$ $\theta_{max} = 6.6$ $N_f =$ 100,110	ω = 1 $m_{max} = 0.54$ $\theta_{max} =5.6$ $N_f = 86,410$	ω = 1 $m_{max} = 0.422$ $\theta_{max} = 4.4$ $N_f = 67,160$	ω = 1 $m_{max} = 0.3$ $\theta_{max} = 3.1$ $N_f =$ 47,750	ω = 1 $m_{max} = 0.218$ $\theta_{max} = 2.2$ $N_f = 34,690$	ω = 1 $m_{max} =0.152$ $\theta_{max} = 1.6$ $N_f =24,190$	ω = 1 $m_{max} = 0.121$ $\theta_{max} = 1.25$ $N_f = 19,250$
X =ω p = 10^7	ω = 1 $m_{max} = 0.629$ $\theta_{max} = 6.6$ $N_f =$ 1,001,100	ω = 1 $m_{max} = 0.54$ $\theta_{max} = 5.6$ $N_f =$ 864,100	ω = 1 $m_{max} = 0.422$ $\theta_{max} = 4.4$ $N_f =$ 671,600	ω = 1 $m_{max} = 0.3$ $\theta_{max} = 3.1$ $N_f =$ 477,500	ω = 1 $m_{max} =0.218$ $\theta_{max} = 2.2$ $N_f =$ 346,900	ω = 1 $m_{max} =0.152$ $\theta_{max} = 1.6$ $N_f =$ 241,900	ω = 1 $m_{max} =0.121$ $\theta_{max} = 1.25$ $N_f =$ 192,500

The data of Table 4.9 obtained theoretically allows the construction of various graphical representations of dependencies that are very important in engineering practice in the design and calculation of composites and nanocomposites for creep-fatigue. Namely: the S – N_f fatigue curve; fatigue curve in the form of damage function v.s. maximum dimensionless temperature θ; fatigue stress (strain)—time curve until complete failure of the structural element. Some of them are presented below.

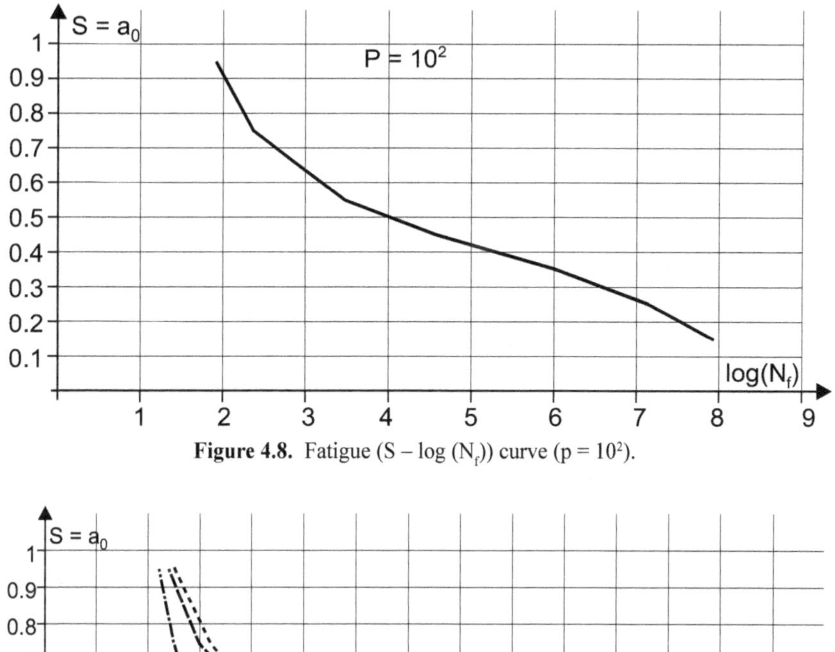

Figure 4.8. Fatigue (S – log (N_f)) curve (p = 10^2).

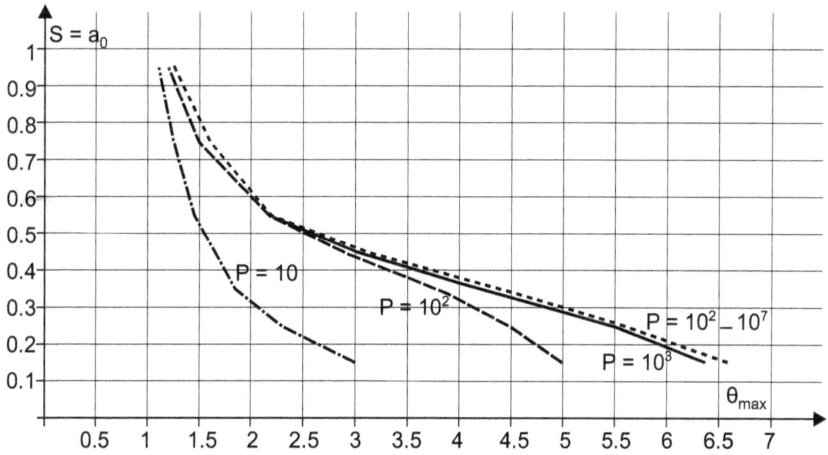

Figure 4.9. Fatigue (S – θ_{max}) curves [a_0 v.s. θ_{max} for different frequencies log(p)].

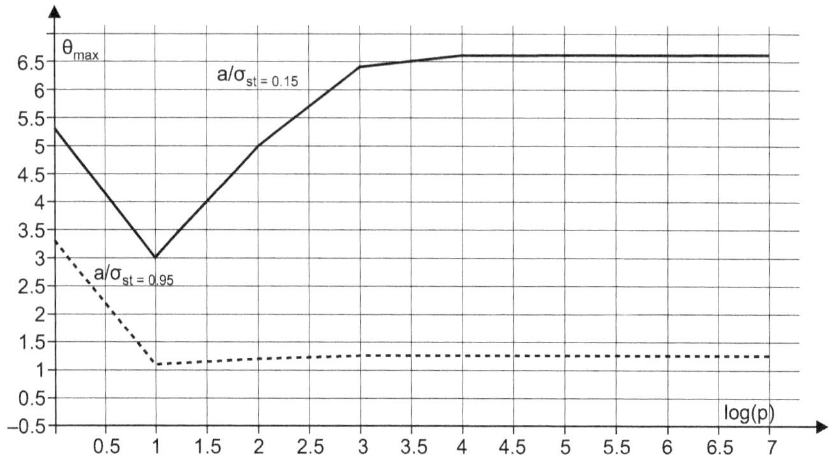

Figure 4.10. Fatigue (θ_{max} – log (p)) curves.

One can see from Fig. 4.10 that the dawning portion of stress (dimensionless stress amplitude of cyclic load)/log (p) diagram is more profound when $a_0 = a/\sigma_{st}$ is low. Obviously, if static load is increased, say σ_{st} is 2 σ_{st}, then the fatigue life should decrease.

Fully reversed fatigue loading

Example 4.9—same as Example 4.1, but $\sigma_{st} = 0$

Model: X = a1*t + a2*t^2 + a3*t^3 + a4*t^4 + a5*t^5

Variable	Value
a1	0.182985
a2	–0.2746896
a3	0.1380661
a4	–0.027453
a5	0.0018769

$$X = \omega = 0.183\theta - 0.275\theta^2 + 0.1386\theta^3 - 0.0274\theta^4 + 1.88(10^{-3})\,\theta^5 \qquad (4.12)$$

$N_f = 10.82 = m_{max}p/2\pi = 0.68(100)/2\pi = 10.82$ and $T_f = 1008[^0K]$ ($T_f = \beta T_*\,\theta_{max} + 600$; $\beta T_* = 0.1\ (600)$); $\theta_{max} = 6.8$ Compare with Example 4.1

$N_f = 7.64 < 10.82$ and $T_f = 1008[^0K] > 780\ [^0K]$ ($T_f = \beta T_*\,\theta_{max} + 600$; $\beta T_* = 0.1\ (600)$);

$\Theta_{max,f} = 5.0 < 6.8$

a1*t + a2*t^2 + a3*t^3 + a4*t^4 + a5*t^5

Variable	Value
a1	0.190983
a2	–0.2610355
a3	0.1173138
a4	-0.020982
a5	0.0012936

$$X = \omega = 0.191\theta - 0.261\theta^2 + 0.117\theta^3 - 0.021\theta^4 + 1.29(10^{-3})\,\theta^5 \tag{4.13}$$

$N_f = 12.1 = m_{max}p/2\pi = 0.76(100)/2\pi = 12.1$ and $T_f = 1056[^0K]$ ($T_f = \beta T_* \theta_{max} + 600$; $\beta T_* = 0.1\ (600)$); $\theta_{max} = 7.6$

Compare with Example 4.9

$N_f = 12.1 \approx 10.82$ and $T_f = 1056[^0K]$ ($T_f = \beta T_* \theta_{max} + 600$; $\beta T_* = 0.1\ (600)$);

The fatigue life is approximately the same and the maximum temperature at failure is:

$\theta_{max} = 7.6 \approx 6.8$ ($T_f = 1056[^0K] \approx T_f = 1008[^0K]$). Therefore, the effect of glass transition temperatures of the matrix and filler on the effective stiffness E and the fatigue life is small. Consider now a fully reversed fatigue loading

Effect of mean load

Well before a microstructural understanding of fatigue processes was developed, engineers had developed empirical means of quantifying the fatigue process and designing such that it is arrested. Perhaps the most important concept is the S – N diagram, such as those shown in Fig. 4.1, in which a constant cyclic stress amplitude S is applied to a specimen and the number of loading cycles N until the specimen fails is determined. Millions of cycles might be required to cause failure at lower loading levels, so the abscissa is usually plotted logarithmically.

Statistical variability is troublesome in fatigue testing; it is necessary to measure the lifetimes of perhaps twenty specimens at each of ten or so load levels to construct the S – N curve with statistical confidence. It is generally impossible to cycle the specimen at more than approximately 10Hz (inertia in components of the testing machine and heating of the specimen often become problematic at higher speeds) and at that speed, it takes 11.6 days to reach 10^7 loading cycles . Obtaining a full S – N curve is obviously a tedious and expensive procedure.

A very substantial amount of testing is required to obtain a S – N curve for the simple case of fully reversed loading, and it will usually be impractical to determine whole families of curves for every combination of mean and alternating stress. A very substantial amount of testing is required to obtain a S – N curve for the simple case of fully reversed loading, and it will usually be impractical to determine whole families of curves for every combination of mean and alternating stress. There are a number of strata gems and one common one being the Goodman diagram (see Fig. 4.7). Here a graph constructed with mean stress as the abscissa and alternating stress as the ordinate, and a straight "lifeline" is drawn from σ_e on the σ_{alt} axis to the ultimate tensile stress σ_f on the σ_m axis. Then for any given mean stress, the endurance limit | the value of alternating stress at which fatigue fracture never occurs | can be read directly as the ordinate of the lifeline at that value of σ_m. Alternatively, if the design application dictates a given ratio of σ_e to σ_{alt}, a line is drawn from the origin with a slope equal to that ratio. Its intersection with the life line then gives the effective endurance limit for that combination of σ_a and σ_m.

The maximum completely reversing cyclic stress that a material can withstand for an indefinite (or infinite) number of stress reversals is known as the fatigue strength or endurance strength (S_e) of the material.

Model: X = a1*t + a2*t^2 + a3*t^3 + a4*t^4 + a5*t^5

Variable	Value
a1	0.1837596
a2	–0.2569729
a3	0.1169506
a4	–0.0211128
a5	0.001312

$$X = \omega = 0.184\theta - 0.257\theta^2 + 0.117\theta^3 - 0.021\theta^4 + 1.31(10^{-3})\,\theta^5 \tag{4.14}$$

$N_f = 12.1 = m_{max}p/2\pi = 0.76(100)/2\pi = 12.1$ and $T_f = 1056[^0K]$ ($T_f = \beta T_* \theta_{max} + 600$; $\beta T_* = 0.1\,(600)$); $\theta_{max} = 7.6$

Compare with Example 4.9

$N_f = 12.1 \approx 10.82$ and $T_f = 1056[^0K]$ ($T_f = \beta T_* \theta_{max} + 600$; $\beta\theta_* = 0.1\,(600)$);

The fatigue life is approximately the same and the maximum dimensionless temperature at failure is:

$\theta_{max} = 7.6 \approx 6.8$ ($T_f = 1056[^0K] \approx T_f = 1008[^0K]$). Therefore, the effect of effective stiffness (mixture law) of the matrix and filler on the effective fatigue life is small. Stiffness change has a drastic effect on the fatigue life only if the filler concentration is high:

fi = $\boldsymbol{\varphi}$ = 0.5. For the example above k = 2 and $\boldsymbol{\theta}_{max}$ = 6.5. If k = 10 then $\boldsymbol{\theta}_{max}$ = 3.0 (see Example 4.11 below).

Example 4.11—The same as Example 4.1 but k =10

Differential equations

1 d(z11)/d(t) = (1/(1-X)^6)*(exp(t/(1+0.067*t)))*A1*((exp(0.1*m)))*m1*(sin (p*m))*z^n

2 d(z12)/d(t) = (1/(1-X)^6)*(exp(t/(1+0.067*t)))*A1*((exp(0.1*m)))*m1*(cos (p*m))*z^n

3 d(z21)/d(t) = (1/(1-X)^6)*(exp(t/(1+0.067*t)))*(A2)*(z^s)*((exp(0.1*m)))*m1*(sin (p*m))*z^n

4 d(z22)/d(t) = (1/(1-X)^6)*(exp(t/(1+0.067*t)))*(A2)*(z^s)*((exp(0.1*m)))*m1*(cos (p*m))*z^n

5 d(z31)/d(t) = (1/(1-X)^2)*(1/k)*(exp(t/(1+0.067*t)))*A3*(z^s)*f36*((exp(0.1*m2)))*m21*((sin (p*m2))^1)*z^n

6 d(z32)/d(t) = (1/(1-X)^2)*(1/k)*(exp(t/(1+0.067*t)))*A3*f36*(z^s)*((exp(0.1*m2)))*m21*1*(cos (p*m2))*z^n

7 d(X1)/d(t) = m1*A*(z^2)/((1-X1)^6)

8 d(X2)/d(t) = m1*A0*(z^2)/((1-X2)^2)

Comparing Examples 4.12 and 4.13 one can conclude that a drastic increase in effective stiffness ratio (k = 10 instead of k = 2) results in a large decrease in fatigue life from θ_{max} = 6.5 to θ_{max} = 2.0. In terms of maximum cyclic creep-fatigue the number of cycles before failure of composite and nanocomposite materials is: $N_f = 0.65(100)/2\pi = 10.34$ to $N_f = 0.2(100)/2\pi = 3.18$ (p = 100; and fi = φ = 0.5).

Cyclic creep-fatigue and hold time case ("Stop and Go")

Example 4.13—The same as Example 4.12 but Y13 instead of Y1

Calculated values of DEQ variables

	Variable	Initial value	Minimal value	Maximal value	Final value
1	a	0.15	0.15	0.15	0.15
2	A	2.	2.	2.	2.
3	A0	2.	2.	2.	2.
4	a1	0.25	0.25	0.25	0.25

5	A1	1.	0.0654108	1.	0.0654108
6	A2	0	0	0.0156653	0.0041085
7	A3	0	0	0.540653	0.540653
8	c1	3.	3.	3.	3.
9	c2	5.	5.	5.	5.
10	d	0	0	6.81767	6.81767
11	deff	0	0	0.1079849	0.1079849
12	deff1	0	0	0.540653	0.540653
13	E	1.448238	0.3625	1.448238	0.3625
14	f3	1.	1.	1.	1.
15	f31	0	0	0.999985	0.9814282
16	f32	1.	0.0654108	1.	0.0654108
17	f33	0	0	0.9997489	0.6438535
18	f34	0	0	0.7262598	0.7262598
19	f35	1.	1.	1.	1.
20	f36	0	0	1.	1.
21	f37	0	0	0	0
22	fi	0.05	0.05	0.05	0.05
23	k	10.	10.	10.	10.
24	m	0	0	0.681767	0.681767
25	m1	0.1	0.1	0.1	0.1
26	m2	0	0	0.0825473	0.0823989
27	m21	0.0868	0.0007285	0.0868	0.0062005
28	n	1.	1.	1.	1.
29	p	100.	100.	100.	100.
30	s	1.	1.	1.	1.
31	t	0	0	7.	7.
32	t1	1.	1.	1.	1.
33	t2	2.	2.	2.	2.
34	X	0	0	0.9772713	0.9772713
35	X1	0	0	0.9485852	0.9485852
36	X2	0	0	0.0286861	0.0286861
37	Y1	0	−0.2079093	0.2140028	−0.0438575
38	Y13	0.2172357	0	0.4333589	0.0868714

39	Y2	0	−0.0862574	0.090621	0
40	z	0.3922357	0.1862823	0.9743384	0.9743384
41	z1	0	−0.7315188	0	−0.7315188
42	z11	0	−8.54486	0.2303334	−8.54486
43	z12	0	−0.3687132	5.355513	5.355513
44	z2	0	−0.0301275	0	−0.0301275
45	z21	0	−0.5091421	0.0070055	−0.5091421
46	z22	0	−0.0123609	0.3357937	0.3357937
47	z3	0	−0.0012073	0	−0.0010252
48	z31	0	−9.157E-05	0.0163475	0.0163475
49	z32	0	−0.0055352	0.0003551	−0.0055352

Differential equations

1 d(z11)/d(t) = (1/(1-X)^6)*(exp(t/(1+0.067*t)))*A1*((exp(0.1*m)))*m1*(sin (p*m))*z^n

2 d(z12)/d(t) = (1/(1-X)^6)*(exp(t/(1+0.067*t)))*A1*((exp(0.1*m)))*m1*(cos (p*m))*z^n

3 d(z21)/d(t) = (1/(1-X)^6)*(exp(t/(1+0.067*t)))*(A2)*(z^s)*((exp(0.1*m)))*m1*(sin (p*m))*z^n

4 d(z22)/d(t) = (1/(1-X)^6)*(exp(t/(1+0.067*t)))*(A2)*(z^s)*((exp(0.1*m)))*m1*(cos (p*m))*z^n

5 d(z31)/d(t) = (1/(1-X)^2)*(1/k)*(exp(t/(1+0.067*t)))*A3*(z^s)*f3*((exp(0.1*m2)))*m21*((sin (p*m2))^1)*z^n

6 d(z32)/d(t) = (1/(1-X)^2)*(1/k)*(exp(t/(1+0.067*t)))*A3*f3*(z^s)*((exp(0.1*m2)))*m21*1*(cos (p*m2))*z^n

7 d(X1)/d(t) = m1*A*(z^2)/((1-X1)^6)

8 d(X2)/d(t) = m21*A0*(z^2)/((1-X2)^2)

Explicit equations

1 m = 0.1*t

2 m1 = 0.1

3 m2 = 0.0868*t - 0.0386*(t^2)+ 0.0080*(t^3) - 0.000756*(t^4) + 2.6*(10^-5)*(t^5)

4 p = (10^2)

5 fi = 0.05

6 c1 = 3*(2-exp(-0.4*t))^0

```
c2 = 5*(2-exp(-0.4*t))^0
t2 = 2
t1 = 1
k = 10
n = 1.0
s = 1.0
E = (0.625-0.375*(tanh(c1*(t-t1))))*(1-fi)+(k)*(0.625-0.375*
(tanh(c2*(t-t2))))*(fi)
z1 = (cos(p*m))*z11-(sin(p*m))*z12
a = 0.15
z2 = (cos(p*m))*z21-(sin(p*m))*z22
a1 = 0.25
Y2 = if (t<4) then (0) else (if (t <6) then (a1)*(sin (p*m))*E else (0))
Y1 = if (t<0) then (0) else ((a)*(sin (p*m))*E)
Y13 = If (t<5) Then ((a)*(1+1*sin (p*t))*E) Else (If (t<6) Then (0) Else
(If (t<7) Then ((a1)*(1+1*sin (p*t))*E) Else (If (t < 10) Then (0) Else
(If (t <12) Then ((a1)*(1+sin (p*t))*E) Else (0)))))
A1 = 1*(exp(-0.4*t))
deff1 = (1+fi*((10*t+1)^0.333-1))^3 - 1
d = 1*t
A2 = 0.1*(exp(-0.4*t))*((1+2.5*fi+14.7*(fi^2)))*deff1^1
deff = (fi*d)/(1 + 0.333*(1 - fi)*d)
z3 = (cos(p*m2))*z31-(sin(p*m2))*z32
X = X1+X2
A = 2
A0 = 2
f3 = if (t<0) then (0) else (1)
f31 = (1/36)*t*(12-t)
f32 = (exp(-0.4*t))
f33 = (sin(3.14*t/4))^2
f34 = (1/64)*t^2
f35 = if (t<7) then (1) else (0)
f36 = if (t<6) then (0) else (1)
f37 = if (t<0) then (1) else (0)
z = Y13-(z1+z2)*((exp(-0.1*m)))-z3*(exp(-0.1*m2))+0.175*1
```

39 m21 = 0.0868 - 0.0772*t + 0.024*(t^2) - 0.003*(t^3) + 1.3*(10^-4)*(t^4)

40 A3 = 1*deffl

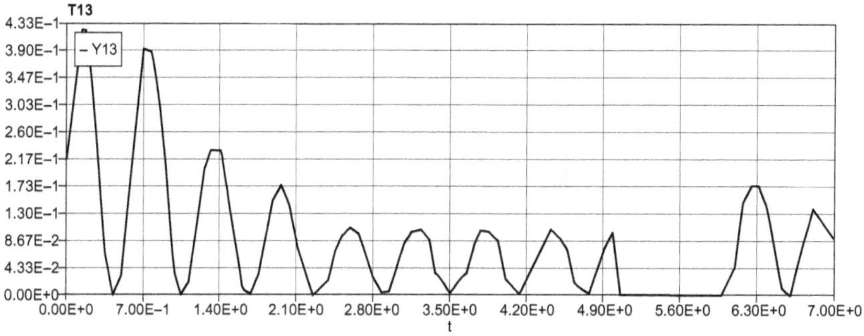

Figure 4.11. Instantaneous stress function Y_{13} (stop and go case).

k = 2 and Y13 = If (t<5) Then ((a)*(1+1*sin (p*t))*E) Else (If (t<6) Then (0) Else (If (t<7) Then ((a1)*(1+1*sin (p*t))*E) Else (If (t < 10) Then (0) Else (If (t <12) Then ((a1)*(1+sin (p*t))*E) Else (0))))). See Fig. 4.12.

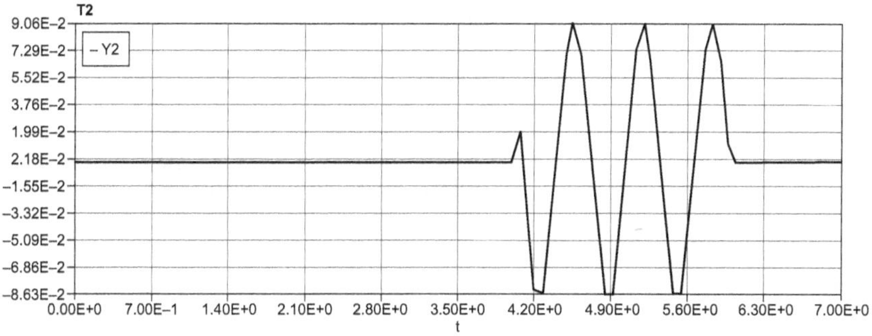

Figure 4.12. Y2 = if (t<4) then (0) else (if (t <6) then (a1)*(sin (p*m))*E else (0)).

References

[1] Harris B. Fatigue in composite. CRC/WP; vol. 3, 2003.

[2] Nicholas T. Critical issues in high cycle fatigue // Int. J. of Fatigue, 21: 221 –231, 1999.

[3] Ladeve`ze P, Lubineau G. A computational meso-damage model for life prediction for laminates. CRC/WP, 3: 432– 41, 2003.

[4] Reifsneider K. L. Fatigue of Composite Materials; Elsevier: New York, NY, USA, 1990,

[5] Yang J. N. Fatigue and residual strength degradation for graphite/epoxy composites under tension-compression cyclic loading. J. Compos. Mater., 12: 19–39, 1978.

[6] Sendeckyj G. P. Fitting models to composite materials fatigue data. pp. 245–260. *In*: Chamis C. C. (Ed.). Test Methods and Design Allowables for Fibrous Composites; ASTM STP 734; ASTM: Philadelphia, PA, USA, 1981.

[7] Wolff R. V. and Lemon G. H. Reliability Predictions for Adhesive Bonds; Air Force Materials Laboratory: Dayton, OH, USA, 1972.
[8] Sevenois R. D. B. and Van Paepegem W. Fatigue damage modeling techniques for textile composites: Review and comparison with unidirectional composites modeling techniques. ASME Appl. Mech. Rev., 67: 1–12, 2015.
[9] D'Amore A. and Grassia L. Phenomenological approach to the study of Hierarchical damage mechanisms in composite materials subjected to fatigue loadings. Comp. Struct., 175: 1–6, 2017.
[10] Vassilopoulos A. P. and Keller T. Fatigue of Fiber-Reinforced Composites; Springer: London, UK, 2013.
[11] Reifsneider, K.L. Fatigue of Composite Materials; Elsevier: New York, NY, USA, 1990.
[12] Razdolsky L. Probability-Based Structural Fire Load, Cambridge University Press, 2014.
[13] Razdolsky L. Creep-fatigue Models of Composites and Nanocomposites (Deterministic and Probabilistic Approach), AIAA – 2019.
[14] Geoffry Laufersky and Thomas Nann. Physical Chemistry of Nanoparticle Syntheses in Comprehensive Nanoscience and Nanotechnology (Second Edition), Wellington, Elsevier, New Zealand, 2019.
[15] Ladeve`ze P. and Lubineau G. A computational meso-damage model for life prediction for laminates. CRC/WP, 3, 2003.
[16] Sevenois R. D. B. and Van Paepegem W. Fatigue damage modeling techniques for textile composites: Review and comparison with unidirectional composites modeling techniques. ASME Appl. Mech., 2015.
[17] D'Amore A. and Grassia L. Phenomenological approach to the study of Hierarchical damage mechanisms in composite materials subjected to fatigue loadings. Comp. Struct., 175: 1–6, 2017.
[18] Vassilopoulos A. P. and Keller T. Fatigue of Fiber-Reinforced Composites; Springer: London, UK, 2013.
[19] Halpin J. C., Jerina K. L. and Johnson T. A. Characterization of composites for the purpose of reliability evaluation. In Analysis of Test Methods for High Modulus Fibers and Composites; ASTM STP: West Conshohocken, PA, USA, pp. 5–64, 1973.
[20] Razdolsky L. Phenomenological Creep Models of Composites and Nanomaterials Deterministic and Probabilistic Approach, CRC Press Taylor & Francis Group, Boca Raton, FL, 2019.
[21] Razdolsky L. "Phenomenological Creep Models of Composites and Nanomaterials (AIAA – SPACE X, American Institute of Aeronautics and Astronautics), 2019.
[22] Bazant Z. P. Why continuum damage is nonlocal: micromechanical arguments. J. Engrg. Mech. 117(5): 1070–1087, 1991.
[23] Bazant Z. P. and Pijaudier-Cabot G. Nonlocal continuum damage, localization instability and convergence. J. Appl. Mech. 55: 287–293, 1988.
[24] Whitney J. M. and Nuismer R. J. Stress fracture criteria for laminated composite containing stress concentration. J. Compos. Mater., 8: 253–6, 1974.
[25] Yang J. N. and Jones D. L. Load Sequence Effects on Graphite/Epoxy [±35]2S Laminate; ASTM STP: West Conshohocken, PA, USA, 813: 246–262, 1983.
[26] Kassapoglou C. Fatigue Life Prediction of Composite Structures Under Constant Amplitude Loading. J. Compos. Mater., 41: 2737–2754, 2007.
[27] Chaboche. Continuum damage mechanics I: General concepts & II: Damage growth, crack Initiation and crack growth. J. Appl. Mech., 55: 59–72, 1988.

CHAPTER 5

Probabilistic Approach to Creep-Fatigue Models

5.1 Introduction

A uniaxial phenomenological energy type model is proposed for describing the inelastic deformation and failure of composites and nanocomposites under the combined action of static and cyclic thermal loads. The amplitude value of the cyclic stress component in the examples below was no more than 95% of the static stress. The Gibbs energy [1,2] and adjacent thermodynamic [3,4] approaches for describing inelastic rheological deformation and failure of composites and nanocomposites under unsteady loading provides good results, and the expediency of their use in computational practice. The purpose of this chapter is to generalize the approach to describe the class of phenomena that occur in the material under the combined action of static σ_{st} and cyclic loads with an amplitude value of the cyclic component σ_a. Consideration is given to the so-called multi-cycle loading at a frequency $f > 10$ Hz and an amplitude coefficient $a_0 = \sigma_a/\sigma_{st}$ not exceeding a certain critical value a_{cr}, having an order of 0.95. In this case, cyclic loading leads to two main effects [5,6]: (1) acceleration (or even initiation) of the creep process at a given static stress σ_{st}; (2) a decrease in the accumulated inelastic deformation at the time of failure compared with a similar value under pure static loading. As a rule, these phenomena cannot be described either within the framework of just ordinary classical approaches, or from the standpoint of phenomenological creep-fatigue only [7]; or from the standpoint of fatigue in a symmetric (or asymmetric) cycle [8].

5.2 Creep-fatigue process under periodic loads

Dr. Talreja [9] correctly described the multifaceted complexity of the problem of cyclic creep-fatigue of composites and nanocomposites: “A reliable and cost-effective fatigue life prediction methodology for composite structures requires a physically based modeling of fatigue damage evolution. An undesirable alternative is an empirical approach. A major obstacle to developing mechanistic models for composites is the complexity of the fatigue damage mechanisms, both in their geometry and the details of the evolution process. Overcoming this obstacle requires insightful simplification that allows the use of well-developed mechanical modeling tools without compromising the essential physical nature of the fatigue process”. Due to the variety of structures of composite materials and the mechanisms of damage accumulation under cyclic loading, it is impossible to construct a universal theory of the fatigue of composites [10]. Although the fatigue behavior of composite materials is significantly different from the behavior of metals, many models have been developed based on the well-known S – N curves. These models make up the first class of so-called “fatigue strength models”. This approach requires large experimental studies and does not take into account the real mechanisms of damage, such as matrix damage and fiber breaks [11]. The second class includes phenomenological models for multi-cycle creep-fatigue. These models offer an evolutionary law that describes the gradual degradation of the strength or stiffness of a composite sample based on macroscopic properties. Recently, models have been developed based on the concepts of continuum damage mechanics. The development of damage is determined by evolutionary kinetic equations that reflect the irreversible nature of damage [12]. Continuum damage models introduce scalar, vector, or tensor damage parameters that describe the degradation of the entire composite material or structural components. These models are based on physical modeling of the main damage mechanisms that lead to macroscopically noticeable degradation of mechanical properties [13]. The main result of all fatigue models is the prediction of fatigue life, and each of these three categories uses its own criterion to determine the fatigue failure process and, as a consequence, the fatigue life of the composite material.

5.3 Continuum damage mechanics and durability of composites

In the continuum approach to the analysis of the stress state and fatigue of composites, the material is considered as a homogeneous anisotropic elastic medium [14]. When constructing the model, it is assumed that small elastic strains take place. The elastic strain energy function is quadratic and there is a linear relationship between the stress tensor σ and the strain tensor ε. The proposed model is based on the concept of effective stress that integrally reflects various types of damage at the micro-scale level (such as the formation and growth of micro cracks in the matrix, fiber breaks, delaminating and other microscopic defects) [15].

5.4 Damage function ω and decrease of cross section area

Damage parameter ω is associated with any crossectional area in the vicinity of a given point on the body and is determined by the following formula:

$$\omega = \frac{A - \Delta A}{A} \tag{5.1}$$

In Equation (5.1) "A" is the nominal, undamaged cross-sectional area, "ΔA" is the total cross-sectional area of all defects in a given structural element. By definition, the theoretical value of ω should be in the range $0 \leq \omega \leq 1$. The effective stress tensor in the case of isotropic damage is introduced as follows:

$$\tilde{\sigma} = \frac{\sigma_{ij}}{1 - \omega} \tag{5.2}$$

Formula (5.2) allows us to take into account the directed, anisotropic nature of the accumulation of fatigue defects. Models of anisotropic damage are significantly more complicated than theoretically isotropic damage, while ensuring compatibility with the thermodynamic principles of continuum mechanics. To identify the parameters of the models of anisotropic damage, a significant number of experiments are required with complex test programs that make it possible to identify the directional nature of fatigue damage. In considering the geometric interpretation of the anisotropic damage function, the second rank tensor is introduced (for physical reasons, it is symmetrical). Generalization of dependence Equation (5.2) to the case of the damage tensor of the second rank, the symmetric shape of the effective stresses σ_{ij} is obtained as follows [16]:

$$\tilde{\sigma}_{ij} = A_1 \frac{[\sigma_{ij}]^{n1}}{[1 - \omega]^{m1}} + A_2 \frac{[\sigma_{ij}]^{n2}}{[1 - \omega]^{m2}} \tag{5.3}$$

With the introduction of effective stress, isotropic damage and using the principle of deformation equivalence [15], the uniaxial integral type constitutive creep equation has a form in dimensionless variables and parameters:

$$E_i(\theta)[\theta] = \sigma_i(\theta) + \int_0^\theta e^{\frac{\tau}{1+\beta\tau}} \frac{1}{[1-\omega]^m} K_i(\theta,\tau)\sigma^n(\tau)\, m\mathbf{1} d\tau$$

$$K_i(\theta,\tau) = \varphi_i(\theta) f_i(\tau) = m\mathbf{1}(\tau) \sum_{j=1}^{N} \exp(-\alpha_j m(\theta)) \exp(\alpha_j m(\tau))$$

$$\frac{d\omega}{d\theta} = \frac{\sigma_i^n}{[1-\omega]^k} m\mathbf{1}(\theta) \quad \omega(\mathbf{0}) = \mathbf{0}; \;\; \omega(\mathbf{0}_f) = \mathbf{1} \tag{5.4}$$

$$i = \mathbf{1,2,3}..., n; \; \beta = \frac{RT_*}{E_a}; \;\; T = \beta T_*\theta + T_*[^0K];$$

$T_*[^0K]$ - Base temperature

The introduction of a scalar measure of damage also determines the choice of a mathematical model for describing the damage accumulation . Under conditions of a high-temperature stress state, the rate of damage accumulation should depend on the joint variants of stress tensors and tensors characterizing the mechanical properties of the composite (the parameters n and k in Equation (5.4) that are determined experimentally). The desire to reflect the characteristics of the cycles for each component of the stress tensor leads to a complication of theoretical models. A realistic approach to such situations is to introduce the amplitude of the instantaneous stress a_0 in Equation (5.4) and frequency p. The scalar measure of damage, ω; considered as a function that depends on the maximum stress value per cycle; the number of loading cycles N; the cycle parameter R; temperature T; material properties and other variables affecting material fatigue. In the proposed model it has been assumed that the damage accumulation rate depends on current stress level and all required material constants are determined from relatively simple experiments at fixed values of the cycle parameter and temperature. The rate of damage level ω is presented as:

$$\frac{d\omega}{d\theta} = f(\sigma, \omega, n_i, m_i) \quad i = \mathbf{1;2} \tag{5.5}$$

In accordance with the concept of continuum damage mechanics, function f(ω) can be obtained theoretically based on the number of loading cycles or by analyzing changes in the modulus of elasticity with the temperature rise. However, for the practical use of the theory, it is more preferable to identify the functional dependence (5.5) based on the S – N Weller fatigue curves' results [16].

The geometric interpretation of the damage parameter at $\omega = 1$ corresponds to the case when the cross section of the composite material is completely filled with macro-cracks. In practice, the material becomes unstable and collapses when the damage reaches a certain critical value, less than unity. Due to significant nonlinearity of the dependence of the damage parameter on the number of cycles, at the stage preceding the failure, the growth rate increases and tends to infinity. Therefore, the interval of change of the damage parameter is close to unity, $0.9 < \omega < 1$ corresponds to an insignificant change in the number of cycles to fatigue life N_f.

The basic experiments to determine the fatigue characteristics of composite materials are experiments on cyclic loading under uniaxial stress conditions. Due to the significant scatter of experimental data, it is necessary to test a large number of samples at various stress levels in order to construct S – N curves. The proposed deterministic and probabilistic models are presented by the examples given below. Moreover, the statistics that are necessary to solve the corresponding probabilistic problems are based on the solutions of the corresponding problems in a deterministic formulation.

5.5 "Forward" and "Reversed" probabilistic problem

The developed model allows the prediction of fatigue strength taking into account the influence of the orientation of the main reinforcement directions relative to the planes of elastic symmetry of the material. The models presented here are using dimensionless parameters and variables that in turn require the minimum number of material parameters in Equation (5.1) and the minimum necessary set of experiments respectfully. A verification analysis using the so-called "reversed probabilistic problem" is performed for composites under various loading conditions. Consider this issue in more detail. The main goal of solving the "forward" problem of cyclic creep and fatigue of a composite (in a deterministic formulation) is to determine the final value of the number of cycles upto failure. Moreover, all parameters and functional dependencies included in the system of Equation (5.4) are considered deterministic and obtained as a result of statistical processing of the corresponding experimental data. The main goal of solving the so-called

"Forward" problem of cyclic creep and fatigue of the composite (in the probabilistic formulation) is to determine the probability of failure (the value of N_f is in a predetermined interval $[N_{f1} \leq N_f \leq N_{f2}]$. In real live practice of structural engineering computations, solving this problem is equivalent to finding the so-called reliability index β (using the so-called FORM (or SORM) method), which, in turn, allows to calculate the probability of a limit state of the system for the entire service life of the structure. The upper bounds of such probabilities (as well as the values of the reliability index β) cannot be calculated mathematically, since they are indicators of the risk that is acceptable to the public in general and engineering practice in particular. The inverse problem in the probabilistic formulation is that the reliability index $\beta_{min/max}$ is given [17]. The parameters and functional dependencies included in the system of Equation (5.4) are considered probabilistic and obtained using the so-called FORM (or SORM) algorithm by the method of successive approximations until the application results of the forward and reverse methods coincide (with a predetermined accuracy). All probability-based computations in this book comply with the correlation theory of random functions and presented below in forms of examples. Probabilistic characteristics (expectation and variance) are calculated under the assumption of the normal distribution law (except in some cases where the use of the Weibull distribution is required). It is also assumed that the total strain of the structural system consists of four components, namely: instant; elastic; viscoelastic and a component characterizing the release (absorption) of heat as a result of the technological process of creating a composite (or heat release as the result of a chemical reaction, for example, the process of creating nanocomposites). Naturally, in order to solve the forward and inverse probabilistic problems it is necessary to establish the influence of the values of individual parameters on the behavior of the system as a whole. Therefore, a large place in this chapter is devoted to the analysis and influence of these parameters and functional dependencies on the result (the construction of S – N fatigue curves analytically). Note that in each case additional necessary explanations are provided, and the results are summarized and presented in form of tables and graphs.

5.6 Phenomenological models of creep-fatigue of composites

The main purpose of this research is the development of the phenomenological model of the cyclic creep-fatigue of composite and nanocomposite materials; to characterize and to take into account the parameters involved in the fatigue-creep process. The specific objectives include the following:

1. The establishment of a phenomenological fatigue life prediction methodology combining the research findings from the previous phenomenological creep constitutive equations (done by the author [15]) with the effects from the specifics of cyclic loading combinations that induce the fatigue failure of composites and nanocomposites at high temperature loading conditions.
2. This phenomenological model should take into account the parameters (both stress-dependent and frequency-dependent) that are involved in the fatigue behavior of the different types of composites and nanocomposites.

Their mathematical presentation is a modified type of regular nonlinear integral creep constitutive equation combined with the so-called time dependent universal damage mechanics of the continuum equation.

3. Analyses of some specific problems associated with the cyclic creep-fatigue process. Particularly relevant is the intensification of creep by high-frequency cyclic loading in composite and nanocomposite materials, which usually occurs at high temperatures, i.e., in the conditions most characteristic of the operation of many critical parts of modern power generating machines. It is also indicative that in composites at high temperatures, cyclic creep develops not only at small values of the amplitude of the cyclic component, but also at values exceeding the static load in some cases.
4. Obtaining analytically quantitative relationships between deformation, cyclic creep rate, temperature and maximum stress of the cycle.
5. Provide analyses of approximate solutions of fatigue-creep constitutive equations of composites.
6. Finally, develop the validation process of the developed models.

In no way rejecting directions in the study of micro-stresses and micro-strains, it should nevertheless be noted that this kind of theory at the level of the mechanics of micro-inhomogeneous media are quite complicated even in the case of a uniaxial stress state and therefore are not well suited for solving, for example, boundary problems of continuum mechanics. For these purposes, conventional phenomenological models of macro-creep and fatigue theory are more preferable. It is clear that the main purpose of the micromechanics of the process of creep-fatigue is to determine the nature of inelastic deformations and justify the choice (at least qualitatively) of material constants and specific functional dependencies. Structural engineers of the future will design special complex structural systems (such as airplanes, rockets and turbines) while simultaneously inventing new materials with the properties that this particular design would require. The idea is to marry the paradigm of 'designing with the known composite material' to a paradigm called 'designing with the new material'. Therefore, it is necessary to know the strength or durability of the composite or nanocomposite. But it is also necessary to know which effect they have, i.e., you need to learn how to manage this process of influence (or optimize the process of structural design). Therefore, the main attention is given to build dependencies between the "main" parameters included in Equation (5.4) and fatigue curves (S – N) or indicators of fatigue life. These dependences and their effects on cyclic creep and fatigue of composites and nanocomposites are given below.

5.7 Failure criteria

Progressive damage models often employ a failure criterion along with degradation models. Different failure criteria have been proposed in the literature such as maximum stress criterion, Haskin's and Puck's criterion.

5.7.1 *Maximum stress theory*

The maximum stress theory predicts failure when the stresses in the principal material axes exceed the corresponding material strength. In order to avoid failure, it has to be ensured that the stress limits are not exceeded

$$-R^c_{\Updownarrow} < \sigma_{\Updownarrow} < R^t_{\Updownarrow}$$

$$-R^c_{\Leftrightarrow} < \sigma_{\Leftrightarrow} < R^t_{\Leftrightarrow} \tag{5.6}$$

$$|\tau|_{\lrcorner} < R_{\lrcorner}$$

As soon as one of the inequalities above is violated the material fails by a failure mode that is associated with the allowable stress. This failure criterion does not take any interaction of stress components into account and therefore certain loading conditions such as superposition of tensile and shear stresses, lead to non-conservative results.

5.7.2 *Haskin's failure theory*

One of the first failure criteria applied in fatigue which distinguished fiber-failure and matrix-failure mode was proposed by Hashin and Rotem. They derived the fatigue failure criterion from their formulation under static loading, which is stated in fiber mode as:

$$\sigma_1 = R^t_{\Updownarrow} \text{ for } \sigma_1 \geq 0$$

$$|\sigma_1| = R^c_{\Updownarrow} \text{ for } \sigma_1 < 0$$

and for the 2D inter – fibre failure:

$$\left(\frac{\sigma_2}{R^t_{\Updownarrow}}\right)^2 + \left(\frac{\tau_{12}}{R_{\lrcorner}}\right)^2 = 1 \text{ for } \sigma_2 \geq 0 \tag{5.7}$$

$$\left(\frac{\sigma_2}{R^c_{\Updownarrow}}\right)^2 + \left(\frac{\tau_{12}}{R_{\lrcorner}}\right)^2 = 1 \text{ for } \sigma_2 < 0$$

In case of inter-fiber failure Hashin and Rotem proposed an elliptic equation, which depends on the transverse stress σ_2 and on the in-plane stress τ_{12}.

In this book the new model of failure of composites and nanocomposites is based on the solution of the equation system (5.7). In this case, it is considered that a structural element (or the system as a whole) has failed if the cumulative damage function $\omega(t)$ takes the value of unity at some point in time (or temperature), in other words, the damage function $\omega(t) = 1$ or in dimensionless form $\omega(\theta) = 1$. The idea of the fracture model $\omega(\theta) = 1$, of course, is not new, but the derivation of this equation, as a rule, was based on purely experimental data or on their combination with the semi-empirical constitutive equation of cyclic creep-fatigue of composites and nanocomposites (see examples below). The advantage of the new model is that there is no need to distinguish part of the failure caused by creep from the part of the failure caused by fatigue of the composite. Another important result of the new model of failure of composites and nanocomposites is that, in addition to obtaining an analytical expression for the S – N fatigue curve, functional dependences $\omega(\theta)$ or $a_0(\theta)$ are also obtained, which can be considered as random processes, which later serve as the basis for constructing a probabilistic approach to solving problems of cyclic creep-fatigue of composites and nanocomposites (see below).

5.8 Specifics of constitutive equation of creep-fatigue of composites

The main characteristic parameters of cyclic loads are as follows: average stress value, cyclic stress amplitude a_0 and frequency p. If the parameters of the cyclic load remain unchanged over time, then this process is called stationary; otherwise—variable. A variety of technological operational modes and the specificity of the properties of composite and nanocomposite materials led to a variety of different phenomenological and physical models. On one hand, the laws of inelastic deformations of such materials and their "lifetime" in the phenomenological formulation are described by one system of fatigue-creep constitutive equations (regardless of the nature of the material itself whether it be metal, wood or nanocomposite) and therefore are very attractive. In addition, it should be noted that the phenomenological models allow obtaining a solution to the problem of composites' fatigue-creep without conducting very expensive tests for fatigue and creep, and limit it to only very simple tests in order to obtain the necessary numerical values of the parameters of the material. In this case, there is no need to take into account the effect of creep and fatigue separately on the result: whether it is effective stress at failure or the lifetime of the material under the action of cyclic loading. On the other hand, when specifying these laws for a new specific material, the question arises about the extrapolation of the computational results beyond the limits of this phenomenological model. Therefore, for reflecting more adequately the process of fatigue creep and inelastic deformation, along with phenomenological theories, the development of theories based on the micro-inhomogeneity of irreversible processes of deformation and chemical transformations under high temperature is necessary. The results of these studies, as a rule, elucidate and complement the originally chosen phenomenological model.

What are the main features of the phenomenological model of fatigue-creep that are different from the corresponding model of pure creep under the action of a high temperature load? First, the instantaneous strain (or stress) should be cyclic. Secondly, the reaction of a material to cyclic strains (stresses) should also be cyclical, and may cause so-called 'beatings' (see examples below), leading to a significant decrease in the life of the composite material. Thirdly, the lifetime of the composite material significantly depends on the oscillation frequency and the value of the initial static stress.

Currently, the problem of structural strength under cyclic loading is considered much more widely. This is due to the development of modern technology in new industries, such as aircraft, power engineering and subsequently the nuclear power industry, chemical engineering and aviation and rocket technology.

It is a well-known fact that in most parts of power machines operating at high temperatures, cyclic loads vary in a wide range of frequencies and amplitudes that are imposed on static loads of various kinds. If the structures operate in a nonstationary time mode, then an uneven temperature field with large gradients arise in its elements, causing considerable cyclic temperature stresses. The cyclic loading significantly reduces the creep resistance in the entire frequency range, and the impact of low frequency loads (tenths and hundredths of a hertz) equal to and exceeding the yield strength of the material reduces fatigue resistance. It became obvious that such traditional characteristics of strength as endurance limit, static creep limits and long-term static strength can no longer suffice the design criteria for reliable performance [18,19]. As a result, new directions appeared in the section of high-temperature strength—cyclical creep and long-term cyclic strength. The issues of structural and surface stability of materials, thermal fatigue and stability of composite structures are also very important in these areas. These circumstances led to the creation of new methods and means of determining the resistance of composites and nanocomposite materials and continuum damage development under cyclic loading. The efforts of many scientists have already achieved significant progress in the field of theoretical interpretations and the quantitative description of the phenomena of cyclic creep and long-term cyclic strength, and in the engineering applications of theoretical results. Particularly relevant is the intensification of creep by high-frequency cyclic loading in composite metallic materials, which usually occurs at high temperatures, i.e., in the conditions most characteristic of the operation of many critical parts of modern power machines. These include working blades and disks of gas turbines and gas pipelines, linings of combustion chambers, fasteners and other parts for which the mode of high-temperature multi-cycle loading is one of the main ones. It is also indicative that in composites at high temperatures, cyclic creep develops not only at small values of the amplitude of the cyclic component, but also at values exceeding the static load in some cases. Most of the known studies in the field of cyclic creep are experimental, and the first of them is only the fact that the dimensionless parameters included in Equation (5.4) affect the final result (fatigues) of the creep of lead, copper, aluminum, and other pure metals.

5.9 Effect of high temperature on fatigue curves ($S - N_f$)

Consider the cyclic fatigue creep behavior of composites and nanocomposites under high temperatures, i.e., building a S – N fatigue curve by an analytical method, provided that the operating temperature – time dependence is specified in one of the four ways: linearly increasing; non-linearly increasing; exponential or logarithmic function. The main prerequisites and methods for constructing fatigue curves were outlined above. It is necessary to emphasize once again that the proposed model is phenomenological, i.e., theoretically suitable for any particular composite or nanocomposite, and therefore

$$
\begin{aligned}
&E[\theta][a_0 \sin(pm)] = \sigma(\theta) + \int_0^{\theta} A_1 e^{\frac{\theta}{1+\beta\theta}} \frac{1}{(1-\omega)^r} K_1(\theta,\tau)\sigma^n(\tau) m\mathbf{1} d\tau + \\
&+\int_0^{\theta} A_2 e^{\frac{\theta}{1+\beta\theta}} \frac{1}{(1-\omega)^r} f_2[\sigma(\tau)] K_2(\theta,\tau)\sigma^n(\tau) m\mathbf{1} d\tau + \\
&+\int_0^{\theta} \frac{E_0}{E_1} A_3 e^{\frac{\theta}{1+\beta\theta}} \frac{1}{(1-\omega)^r} f_3[d(\tau)] K_3(\theta,\tau)\sigma(\tau) m\mathbf{21} d\tau + \sigma_{st} \\
&\frac{d\omega}{d\theta} = A\frac{m\sigma^q}{(1-\omega)^r} + A_0 \frac{m\mathbf{21}\sigma^q}{(1-\omega)^r}; \ \omega(0) = \mathbf{0};\ \omega(\theta_f) = \mathbf{1};\ \sigma(\mathbf{0}) = \mathbf{0} \qquad (5.8) \\
&K_1(\theta,\tau) = \sum_{i=1}^{N} \sin[pm(\theta) - pm(\tau))]\exp(-\alpha_i m\theta)\exp(\alpha_i m\tau) \\
&K_2(\theta,\tau) = \sum_{i=1}^{N} \sin[pm(\theta) - pm(\tau)]\exp(-\beta_{i2} m\theta)\exp(\beta_{i2} m\tau) \\
&K_3(\theta,\tau) = \sum_{i=1}^{N} \sin[pm\mathbf{2}(\theta) - pm\mathbf{2}(\tau)]\exp(-\beta_{i3}\theta)\exp(\beta_{i3} m\mathbf{2})
\end{aligned}
$$

$$E = \left(a - b\left(\tanh\left(c\left(\theta - \theta_{gm}\right)\right)\right)\right)(\mathbf{1} - \varphi) + \left(a - b\left(\tanh\left(\mathbf{c1}\left(\theta - \theta_{gf}\right)\right)\right)\right)\varphi / k$$

$f_2(\sigma) = \sigma^s$; $s = \mathbf{1,2,3}..., N$; $k = E_0 / E_1$; $A_1 = ç = \mathbf{exp}(-\mathbf{b}_1(\beta\theta))$ – Reynolds formula

$A_2 = A_1(\exp(-b_1\beta\theta))((1+2.5\varphi+14.7(\varphi^2))d_{eff1}$;$d=a_1\theta$; $A_3 = [f_3(\theta)][d_{eff1}(\theta)]$;

$d_{eff1} = (1+fi((10t+1)^{1/3}-1))^3 - 1$; $m = 0.01t$; $m1 = 0.01$

$m2 = 0.0868\theta - 0.0386(\theta^2) + 0.0080(\theta^3) - 0.000756(\theta^4) + 2.6(10^{-5})(\theta^5)$

$m21 = 0.0868 - 0.0772\theta + 0.024(\theta^2) - 0.003(\theta^3) + 1.3(10^{-4})(\theta^4)$

$\beta = \dfrac{RT_*}{E_a}$; $T = \beta T_*\theta + T_*$; T_* – Base temperature $[^0K]$

All the physical and physicochemical parameters are predefined (based on the corresponding experimental results). A summary of the calculation and analysis methodology is given above via examples and should be adjusted for each specific case. Examples of the Volterra Integral Equation of the Second Kind and Differential Equation of Cumulative Damage are presented above.

The effect of increasing values of amplitude of mechanical oscillations on maximum fatigue temperature has been illustrated.

It was done before the full output is presented for one case $a_0 = 0.15$. The other solutions for a_o from the interval [0.15….0.95] are presented in short form.

5.10 Examples

Example 5.1—Temperature – time relationships is linearly increasing and a_0 [0.15; 0.25….., 0.95]

Data: $\boldsymbol{\theta} = a\boldsymbol{\tau}$; $\boldsymbol{\tau} = m = [1/a]\,\boldsymbol{\theta}$ and $\boldsymbol{\tau}' = m1 = [1/a]$; $a_0 = 0.15$; $t1 = 4$; $t2 = 6$

Solution of Example 5.1 (a_0 = 0.15) by using POLYMATH software is as follows:

Calculated values of DEQ variables

	Variable	Initial value	Minimal value	Maximal value	Final value
1	A	2.	2.	2.	2.
2	A0	2.	2.	2.	2.
3	A1	1.	0.1470948	1.	0.1470948
4	A2	0	0	0.0156649	0.0077408
5	A3	0	0	0.4529755	0.4529755
6	c1	3.	3.	3.	3.
7	c2	5.	5.	5.	5.

8	d	0	0	4.791695	4.791695
9	deff	0	0	0.09523	0.09523
10	deffl	0	0	0.4529755	0.4529755
11	E	1.45	0.7436083	1.45	0.7436083
12	f3	1.	1.	1.	1.
13	f31	0	0	0.9594444	0.9594444
14	f32	1.	0.1470948	1.	0.1470948
15	f33	0	0	0.999957	0.3374956
16	f34	0	0	0.3587553	0.3587553
17	f35	1.	1.	1.	1.
18	f36	0	0	0	0
19	f37	0	0	0	0
20	fi	0.05	0.05	0.05	0.05
21	k	10.	10.	10.	10.
22	m	0	0	0.4791695	0.4791695
23	m1	0.1	0.1	0.1	0.1
24	m2	0	0	0.0769335	0.0769335
25	m21	0.0868	0.0007284	0.0868	0.0064055
26	n	1.	1.	1.	1.
27	p	100.	100.	100.	100.
28	s	1.	1.	1.	1.
29	t	0	0	5.	5.
30	t1	4.	4.	4.	4.
31	t2	6.	6.	6.	6.
32	X	0	0	0.9815094	0.9815094
33	X1	0	0	0.9432845	0.9432845
34	X2	0	0	0.0382249	0.0382249
35	Y1	0	–0.2162589	0.2174991	–0.0794733
36	z	0.175	–0.0399816	0.5305227	0.5305227
37	z1	0	–0.4480435	0.1616223	–0.4480435
38	z11	0	–5.659615	0.2222582	–5.659615
39	z12	0	–6.202416	0.0040718	–6.202416
40	z2	0	–0.0083036	0.0031161	–0.0083036
41	z21	0	–0.160416	0.0031002	–0.160416

42	z22	0	–0.1696317	2.767E-05	–0.1696317
43	z3	0	0	0	0
44	z31	0	0	0	0
45	z32	0	0	0	0

Differential equations

1 d(z11)/d(t) = (1/(1-X)^6)*(exp(t/(1+0.067*t)))*A1*((exp(0.1*m)))*m1*(sin (p*m))*z^n

2 d(z12)/d(t) = (1/(1-X)^6)*(exp(t/(1+0.067*t)))*A1*((exp(0.1*m)))*m1*(cos (p*m))*z^n

3 d(z21)/d(t) = (1/(1-X)^6)*(exp(t/(1+0.067*t)))*(A2)*(z^s)*((exp(0.1*m)))*m1*(sin (p*m))*z^n

4 d(z22)/d(t) = (1/(1-X)^6)*(exp(t/(1+0.067*t)))*(A2)*(z^s)*((exp(0.1*m)))*m1*(cos (p*m))*z^n

5 d(z31)/d(t) = (1/(1-X)^6)*(1/k)*(exp(t/(1+0.067*t)))*A3*(z^s)*f36*((exp(0.1*m2)))*m21*((sin (p*m2))^1)*z^n

6 d(z32)/d(t) = (1/(1-X)^6)*(1/k)*(exp(t/(1+0.067*t)))*A3*f36*(z^s)*((exp(0.1*m2)))*m21*1*(cos (p*m2))*z^n

7 d(X1)/d(t) = m*A*(z^2)/((1-X1)^6)

8 d(X2)/d(t) = m2*A0*(z^2)/((1-X2)^2)

Explicit equations

1 m = 0.1*t

2 m1 = 0.1

3 m2 = 0.0868*t - 0.0386*(t^2)+ 0.0080*(t^3) - 0.000756*(t^4) + 2.6*(10^-5)*(t^5)

4 p = (10^2)

5 fi = 0.05

6 c1 = 3*(2-exp(-0.4*t))^0

7 c2 = 5*(2-exp(-0.4*t))^0

8 t2 = 6

9 t1 = 4

10 k = 10

11 n = 1.0

12 s = 1.0

13 E = (0.625-0.375*(tanh(c1*(t-t1))))*(1-fi)+(k)*(0.625-0.375*(tanh(c2*(t-t2))))*(fi)

14 z1 = (cos(p*m))*z11-(sin(p*m))*z12

15 z2 = (cos(p*m))*z21-(sin(p*m))*z22

16 Y1 = (0.15)*(sin (p*m))*E

17 A1 = 1*(exp(-0.4*t))

18 deff1 = (1+fi*((10*t+1)^0.333-1))^3 - 1

19 d = 1*t

20 A2 = 0.1*(exp(-0.4*t))*((1+2.5*fi+14.7*(fi^2)))*deff1^1

21 deff = (fi*d)/(1 + 0.333*(1 - fi)*d)

22 z3 = (cos(p*m2))*z31-(sin(p*m2))*z32

23 X = X1+X2

24 A = 2

25 A0 = 2

26 f3 = if (t<0) then (0) else (1)

27 f31 = (1/36)*t*(12-t)

28 f32 = (exp(-0.4*t))

29 f33 = (sin(3.14*t/4))^2

30 f34 = (1/64)*t^2

31 f35 = if (t<7) then (1) else (0)

32 f36 = if (t<6) then (0) else (1)

33 f37 = if (t<0) then (1) else (0)

34 z = Y1-(z1+z2)*((exp(-0.1*m)))-z3*(exp(-0.1*m2))+0.175*1

35 m21 = 0.0868 - 0.0772*t + 0.024*(t^2) - 0.003*(t^3) + 1.3*(10^-4)* (t^4)

36 A3 = 1*deff1

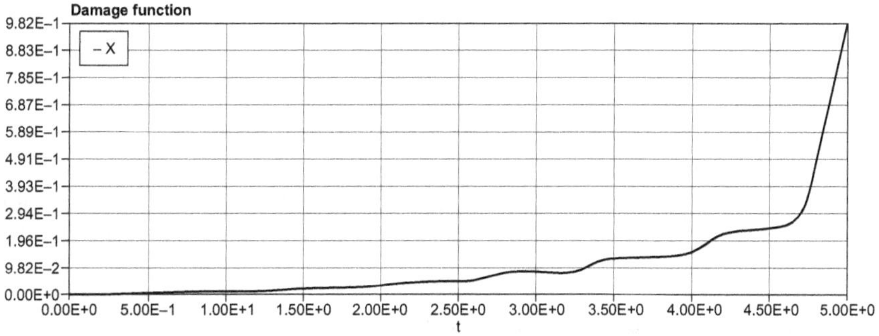

Figure 5.1. Damage function $X = \omega$; $a_0 = 0.15$.

Model: [5.20 – 5.26]

$$X = \omega = -0.04077\theta + 0.0236\theta^2 - 0.00340\,\theta^3 + 1.492(10^{-4})\,\theta^4 \tag{5.19}$$

$N_f = 22.28 = m_{max}p/2\pi = 1.4(100)/2\pi = 22.28$ (it can be also counted from Fig. 4.1) and $T_f = 1440[^0K]$ ($T_f = \beta T_* \theta_{max} + 600$; $\beta T_* = 0.1$ (600)); $\theta_{max} = 14.0$

Example 5.11—same as Example 5.1, but σ_{st} is 0 and $a_0 = 0.25$

Differential equations

1 d(z11)/d(t) = (1/(1-X)^6)*(exp(t/(1+0.067*t)))*A1*((exp(0.1*m)))*m1*(sin (p*m))*z^n
2 d(z12)/d(t) = (1/(1-X)^6)*(exp(t/(1+0.067*t)))*A1*((exp(0.1*m)))*m1*(cos (p*m))*z^n
3 d(z21)/d(t) = (1/(1-X)^6)*(exp(t/(1+0.067*t)))*(A2)*(z^s)*((exp(0.1*m)))*m1*(sin (p*m))*z^n
4 d(z22)/d(t) = (1/(1-X)^6)*(exp(t/(1+0.067*t)))*(A2)*(z^s)*((exp(0.1*m)))*m1*(cos (p*m))*z^n
5 d(z31)/d(t) = (1/(1-X)^6)*(1/k)*(exp(t/(1+0.067*t)))*A3*(z^s)*f3*((exp(0.1*m2)))*m21*((sin (p*m2))^1)*z^n
6 d(z32)/d(t) = (1/(1-X)^6)*(1/k)*(exp(t/(1+0.067*t)))*A3*f3*(z^s)*((exp(0.1*m2)))*m21*1*(cos (p*m2))*z^n
7 d(X1)/d(t) = m*A*(z^2)/((1-X1)^6)
8 d(X2)/d(t) = m2*A0*(z^2)/((1-X2)^2)

Model: X = a1*t + a2*t^2 + a3*t^3 + a4*t^4

Variable	Value
a1	–0.0597141
a2	0.0684915
a3	–0.0175629
a4	0.0013927

$$X = \omega = -0.0597\theta + 0.0685\theta^2 - 0.0176\,\theta^3 + 1.392(10^{-3})\theta^4 \tag{5.20}$$

$N_f = 12.29 = m_{max}p/2\pi = 0.772(100)/2\pi = 12.29$ (it can be also counted from Fig. 5.1 or Table 5.1) and $T_f = 1080[^0K]$ ($T_f = \beta T_* \theta_{max} + 600$; $\beta T_* = 0.1$ (600)); $\theta_{max} = 8.0$

Example 5.12—same as Example 5.1, but σ_{st} is 0 and $a_0 = 0.35$

Differential equations

d(z11)/d(t) = (1/(1-X)^6)*(exp(t/(1+0.067*t)))*A1*((exp(0.1*m)))*m1*(sin (p*m))*z^n

d(z12)/d(t) = (1/(1-X)^6)*(exp(t/(1+0.067*t)))*A1*((exp(0.1*m)))*m1*(cos (p*m))*z^n

d(z21)/d(t) = (1/(1-X)^6)*(exp(t/(1+0.067*t)))*(A2)*(z^s)*((exp(0.1*m)))*m1*(sin (p*m))*z^n

d(z22)/d(t) = (1/(1-X)^6)*(exp(t/(1+0.067*t)))*(A2)*(z^s)*((exp(0.1*m)))*m1*(cos (p*m))*z^n

d(z31)/d(t) = (1/(1-X)^6)*(1/k)*(exp(t/(1+0.067*t)))*A3*(z^s)*f3*((exp(0.1*m2)))*m21*((sin (p*m2))^1)*z^n

d(z32)/d(t) = (1/(1-X)^6)*(1/k)*(exp(t/(1+0.067*t)))*A3*f3*(z^s)*((exp(0.1*m2)))*m21*1*(cos (p*m2))*z^n

d(X1)/d(t) = m*A*(z^2)/((1-X1)^6)

d(X2)/d(t) = m2*A0*(z^2)/((1-X2)^2)

Model: X = a1*t + a2*t^2 + a3*t^3 + a4*t^4

Variable	Value
a1	–0.1245097
a2	0.2067218
a3	–0.0783376
a4	0.0092241

$$X = \omega = -\,0.124\theta + 0.207\theta^2 - 0.0783\,\theta^3 + 9.22(10^{-3})\,\theta^4 \qquad (5.21)$$

$N_f = 7.97 = m_{max}p/2\pi = 0.501(100)/2\pi = 7.97$ (it can be also counted from Fig. 4.1) and $T_f = 912[^0K]$ ($T_f = \beta T_*\,\theta_{max} + 600$; $\beta T_* = 0.1\,(600)$); $\theta_{max} = 5.2$

Example 5.13—same as Example 5.1, but σ_{st} is 0 and $a_0 = 0.45$

Differential equations

1 d(z11)/d(t) = (1/(1-X)^6)*(exp(t/(1+0.067*t)))*A1*((exp(0.1*m)))*m1*(sin (p*m))*z^n

2 d(z12)/d(t) = (1/(1-X)^6)*(exp(t/(1+0.067*t)))*A1*((exp(0.1*m)))*m1*(cos (p*m))*z^n

3 d(z21)/d(t) = (1/(1-X)^6)*(exp(t/(1+0.067*t)))*(A2)*(z^s)*((exp(0.1*m)))*m1*(sin (p*m))*z^n

4 d(z22)/d(t) = (1/(1-X)^6)*(exp(t/(1+0.067*t)))*(A2)*(z^s)*((exp(0.1*m)))*m1*(cos (p*m))*z^n
5 d(z31)/d(t) = (1/(1-X)^6)*(1/k)*(exp(t/(1+0.067*t)))*A3*(z^s)*f3*((exp(0.1*m2)))*m21*((sin (p*m2))^1)*z^n
6 d(z32)/d(t) = (1/(1-X)^6)*(1/k)*(exp(t/(1+0.067*t)))*A3*f3*(z^s)*((exp(0.1*m2)))*m21*1*(cos (p*m2))*z^n
7 d(X1)/d(t) = m*A*(z^2)/((1-X1)^6)
8 d(X2)/d(t) = m2*A0*(z^2)/((1-X2)^2)

Model: X = a1*t + a2*t^2 + a3*t^3 + a4*t^4

Variable	Value
a1	–0.1515534
a2	0.3409575
a3	–0.1709551
a4	0.0281394

$$X = \omega = -\,0.152\theta + 0.341\theta^2 - 0.171\theta^3 + 2.82(10^{-2})\,\theta^4 \qquad (5.22)$$

$N_f = 5.62 = m_{max}p/2\pi = 0.353(100)/2\pi = 5.62$ (it can be also counted from Fig. 4.1) and $T_f = 816[^0K]$ ($T_f = \beta T_* \theta_{max} + 600$; $\beta T_* = 0.1\ (600)$); $\theta_{max} = 3.6$

Example 5.14—same as Example 5.1, but σ_{st} is 0 and $a_0 = 0.55$

Differential equations

1 d(z11)/d(t) = (1/(1-X)^6)*(exp(t/(1+0.067*t)))*A1*((exp(0.1*m)))*m1*(sin (p*m))*z^n
2 d(z12)/d(t) = (1/(1-X)^6)*(exp(t/(1+0.067*t)))*A1*((exp(0.1*m)))*m1*(cos (p*m))*z^n
3 d(z21)/d(t) = (1/(1-X)^6)*(exp(t/(1+0.067*t)))*(A2)*(z^s)*((exp(0.1*m)))*m1*(sin (p*m))*z^n
4 d(z22)/d(t) = (1/(1-X)^6)*(exp(t/(1+0.067*t)))*(A2)*(z^s)*((exp(0.1*m)))*m1*(cos (p*m))*z^n
5 d(z31)/d(t) = (1/(1-X)^6)*(1/k)*(exp(t/(1+0.067*t)))*A3*(z^s)*f3*((exp(0.1*m2)))*m21*((sin (p*m2))^1)*z^n
6 d(z32)/d(t) = (1/(1-X)^6)*(1/k)*(exp(t/(1+0.067*t)))*A3*f3*(z^s)*((exp(0.1*m2)))*m21*1*(cos (p*m2))*z^n
7 d(X1)/d(t) = m*A*(z^2)/((1-X1)^6)
8 d(X2)/d(t) = m2*A0*(z^2)/((1-X2)^2)

Model: X = a1*t + a2*t^2 + a3*t^3 + a4*t^4

Variable	Value
a1	–0.2270152
a2	0.7673331
a3	–0.6396274
a4	0.1756898

$$X = \omega = -0.227\theta + 0.767\theta^2 - 0.639\theta^3 + 1.76(10^{-1})\,\theta^4 \qquad (5.23)$$

$N_f = 3.565 = m_{max}p/2\pi = 0.224(100)/2\pi = 3.565$ (it can be also counted from Fig. 1.1) and $T_f = 738[^0K]$ ($T_f = \beta T_* \theta_{max} + 600$; $\beta T_* = 0.1\ (600)$); $\theta_{max} = 2.3$

Example 5.15—same as Example 5.1, but σ_{st} is 0 and $a_0 = 0.75$

Differential equations

1 d(z11)/d(t) = (1/(1-X)^6)*(exp(t/(1+0.067*t)))*A1*((exp(0.1*m)))*m1*(sin (p*m))*z^n
2 d(z12)/d(t) = (1/(1-X)^6)*(exp(t/(1+0.067*t)))*A1*((exp(0.1*m)))*m1 *(cos (p*m))*z^n
3 d(z21)/d(t) = (1/(1-X)^6)*(exp(t/(1+0.067*t)))*(A2)*(z^s)*((exp(0.1*m)))*m1*(sin (p*m))*z^n
4 d(z22)/d(t) = (1/(1-X)^6)*(exp(t/(1+0.067*t)))*(A2)*(z^s)*((exp(0.1*m)))*m1*(cos (p*m))*z^n
5 d(z31)/d(t) = (1/(1-X)^6)*(1/k)*(exp(t/(1+0.067*t)))*A3*(z^s)*f3*((exp(0.1*m2)))*m21*((sin (p*m2))^1)*z^n
6 d(z32)/d(t) = (1/(1-X)^6)*(1/k)*(exp(t/(1+0.067*t)))*A3*f3*(z^s)*((exp(0.1*m2)))*m21*1*(cos (p*m2))*z^n
7 d(X1)/d(t) = m*A*(z^2)/((1-X1)^6)
8 d(X2)/d(t) = m2*A0*(z^2)/((1-X2)^2)

Model: X = a1*t + a2*t^2 + a3*t^3 + a4*t^4

Variable	Value
a1	–0.3427825
a2	1.741289
a3	–2.226269
a4	0.9210042

$$X = \omega = -0.342\theta + 1.741\theta^2 - 2.226\theta^3 + 9.21(10^{-1})\,\theta^4 \qquad (5.24)$$

$N_f = 2.387 = m_{max}p/2\pi = 0.15(100)/2\pi = 2.387$ (it can be also counted from Fig. 1.1) and $T_f = 690[^0K]$ ($T_f = \beta T_* \theta_{max} + 600$; $\beta T_* = 0.1$ (600)); $\theta_{max} = 1.5$

Example 5.16—same as Example 5.1, but σ_{st} is 0 and $a_0 = 0.95$

Differential equations

1 d(z11)/d(t) = (1/(1-X)^6)*(exp(t/(1+0.067*t)))*A1*((exp(0.1*m)))*m1*(sin (p*m))*z^n

2 d(z12)/d(t) = (1/(1-X)^6)*(exp(t/(1+0.067*t)))*A1*((exp(0.1*m)))*m1*(cos (p*m))*z^n

3 d(z21)/d(t) = (1/(1-X)^6)*(exp(t/(1+0.067*t)))*(A2)*(z^s)*((exp(0.1*m)))*m1*(sin (p*m))*z^n

4 d(z22)/d(t) = (1/(1-X)^6)*(exp(t/(1+0.067*t)))*(A2)*(z^s)*((exp(0.1*m)))*m1*(cos (p*m))*z^n

5 d(z31)/d(t) = (1/(1-X)^6)*(1/k)*(exp(t/(1+0.067*t)))*A3*(z^s)*f3*((exp(0.1*m2)))*m21*((sin (p*m2))^1)*z^n

6 d(z32)/d(t) = (1/(1-X)^6)*(1/k)*(exp(t/(1+0.067*t)))*A3*f3*(z^s)*((exp(0.1*m2)))*m21*1*(cos (p*m2))*z^n

7 d(X1)/d(t) = m*A*(z^2)/((1-X1)^6)

8 d(X2)/d(t) = m2*A0*(z^2)/((1-X2)^2)

Model: X = a1*t + a2*t^2 + a3*t^3 + a4*t^4

Variable	Value
a1	0.0718596
a2	–0.2365555
a3	0.6840137
a4	–0.3121106

$$X = \omega = 0.0718\theta - 0.236\theta^2 + 0.684\theta^3 - 0.312(10^{-1})\,[\theta^4] \quad (5.25)$$

After repeating analogous calculations for different frequencies "p" from the interval $[1 - 10^7]$, can be presented in tabular form (see Table 5.10 below). Now, according to Table 5.10, it is possible to build composite fatigue curves, damage function curves, and so on.

Table 5.10. Maximum dimensionless temperature θ_{max} (frequency $p = 1 - 10^7$).

a/σ_{st}	0.15	0.25	0.35	0.45	0.55	0.75	0.95
X=ω p = 1	$\omega = 1.0$ $m_{max}=1.13$ $\theta_{max}=11.8$ $N_f=0.18$	$\omega = 1.0$ $m_{max}=0.96$ $\theta_{max}=10.0$ $N_f=0.153$	$\omega = 1.0$ $m_{max}=0.844$ $\theta_{max}=8.8$ $N_f=0.134$	$\omega = 1.0$ $m_{max}=$ 0.756 $\theta_{max}=7.8$ $N_f=0.120$	$\omega = 1.0$ $m_{max}=$ 0.682 $\theta_{max}=7.0$ $N_f = 0.108$	$\omega = 1.0$ $m_{max}=0.55$ $\theta_{max}=5.5$ N_f =0.0875	$\omega = 1.0$ $m_{max}=$ 0.462 $\theta_{max}=4.8$ N_f =0.0735
p = 10	$\omega = 1$ $m_{max}=0.76$ $\theta_{max}=8.0$ $N_f=1.216$	$\omega = 1$ $m_{max}=0.578$ $\theta_{max}=6.0$ $N_f=0.92$	$\omega = 1$ $m_{max}=0.271$ $\theta_{max}=2.80$ $N_f=0.431$	$\omega = 1$ $m_{max}=$ 0.196 $\theta_{max}=2.0$ $N_f=0.312$	$\omega = 1$ $m_{max}=$ 0.166 $\theta_{max}=1.7$ $N_f=0.226$	$\omega = 1$ $m_{max}=$ 0.136 $\theta_{max}=1.40$ $N_f = 0.216$	$\omega = 1$ $m_{max}=$ 0.118 $\theta_{max}=1.2$ $N_f=0.170$
X=ω p = 10^2	$\omega = 1.0$ $m_{max}=1.4$ $\theta_{max}=14.0$ $N_f=22.28$	$m_{max}=0.772$ $\theta_{max}= 8.0$ $N_f=12.29$	$\omega = 1.0$ $m_{max}= 0.50$ $\theta_{max}=5.2$ $N_f=7.97$	$\omega = 1.0$ $m_{max}=$ 0.353 $\theta_{max}=3.2$ $N_f=5.62$	$\omega = 1.0$ $m_{max}= 0.22$ $\theta_{max}= 2.2$ $N_f=3.5$	$\omega = 1.0$ $m_{max}=$ 0.15 $\theta_{max}=1.5$ $N_f=2.39$	$\omega = 1.0$ $m_{max}= 0.10$ $\theta_{max}=1.0$ $N_f=1.59$
X=ω p = 10^3	$\omega = 1.0$ $m_{max}=1.66$ $\theta_{max}=17.0$ $N_f=264.2$	$\omega = 1.0$ $m_{max}=$ 0.835 $\theta_{max}=8.6$ $N_f=84.4$	$\omega = 1.0$ $m_{max}=0.522$ $\theta_{max}=5.4$ $N_f=83.09$	$\omega = 1.0$ $m_{max}=$ 0.364 $\theta_{max}= 3.7$ $N_f= 57.93$	$\omega = 1.0$ $m_{max}=$ 0.243 $\theta_{max}=2.55$ $N_f=38.67$	$\omega = 1.0$ $m_{max}=0.16$ $\theta_{max}=1.6$ $N_f=25.5$	$\omega = 1.0$ $m_{max}=$ 0.122 $\theta_{max}=1.25$ $N_f=19.42$
X=ω p = 10^4	$\omega = 1$ $m_{max}=1.65$ $\theta_{max}= 16.6$ $N_f= 26420$	$\omega = 1$ $m_{max}= 0.54$ $\theta_{max}=5.6$ $N_f=864.1$	$\omega = 1$ $m_{max}=0.422$ $\theta_{max}= 4.4$ $N_f=671.6$	$\omega = 1$ $m_{max}= 0.3$ $\theta_{max}= 3.1$ $N_f= 477.5$	$\omega = 1$ $m_{max}=$ 0.218 $\theta_{max}= 2.2$ $N_f=346.9$	$\omega = 1$ $m_{max}=$ 0.152 $\theta_{max}= 1.6$ $N_f=241.9$	$\omega = 1$ $m_{max}=$ 0.121 $\theta_{max}= 1.25$ $N_f=192.6$
X=ω p = 10^5	$\omega = 1$ $m_{max}=16.5$ $\theta_{max}= 16.6$ $N_f=$ 10,011	$\omega = 1$ $m_{max}= 0.54$ $\theta_{max}=5.6$ $N_f=8,641$	$\omega = 1$ $m_{max}= 0.422$ $\theta_{max}= 4.4$ $N_f=6,716$	$\omega = 1$ $m_{max}= 0.3$ $\theta_{max}= 3.1$ $N_f= 4,775$	$\omega = 1$ $m_{max}=$ 0.218 $\theta_{max}= 2.2$ $N_f=3,469$	$\omega = 1$ $m_{max}=$ 0.152 $\theta_{max}= 1.6$ $N_f=2,419$	$\omega = 1$ $m_{max}=$ 0.121 $\theta_{max}= 1.25$ $N_f=1,926$
X=ω p = 10^6	$\omega = 1$ $m_{max}=$ 0.629 $\theta_{max}= 6.6$ $N_f=$ 100,110	$\omega = 1$ $m_{max}= 0.54$ $\theta_{max}=5.6$ $N_f=86,410$	$\omega = 1$ $m_{max}= 0.422$ $\theta_{max}= 4.4$ $N_f=67,160$	$\omega = 1$ $m_{max}= 0.3$ $\theta_{max}= 3.1$ $N_f=$ 47,750	$\omega = 1$ $m_{max}=$ 0.218 $\theta_{max}= 2.2$ $N_f=34,690$	$\omega = 1$ $m_{max}=$ 0.152 $\theta_{max}= 1.6$ $N_f=$ 24,190	$\omega = 1$ $m_{max}=$ 0.121 $\theta_{max}= 1.25$ N_f =19,250
X=ω p = 10^7	$\omega = 1$ $m_{max}=$ 0.629 $\theta_{max}= 6.6$ $N_f=$ 1,001,100	$\omega = 1$ $m_{max}= 0.54$ $\theta_{max}=5.6$ N_f =864,100	$\omega = 1$ $m_{max}=0.422$ $\theta_{max}= 4.4$ $N_f=671,600$	$\omega = 1$ $m_{max}= 0.3$ $\theta_{max}= 3.1$ N_f =477,500	$\omega = 1$ $m_{max}=$ 0.218 $\theta_{max}= 2.2$ N_f =346,900	$\omega = 1$ $m_{max}=$ 0.152 $\theta_{max}= 1.6$ N_f =241,900	$\omega = 1$ $m_{max}=$ 0.121 $\theta_{max}= 1.25$ N_f =192,500

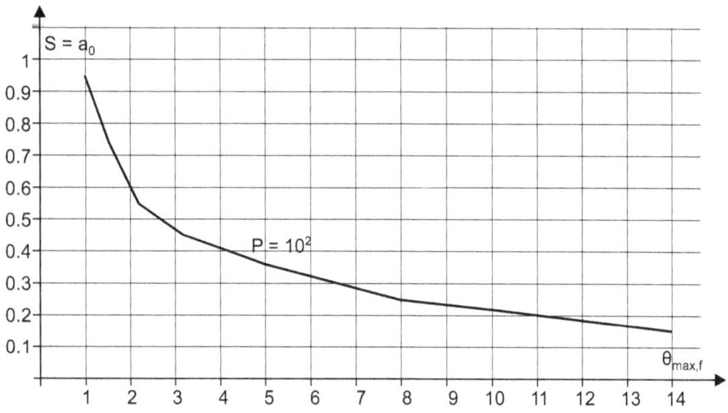

Figure 5.2. $S - N_f$ fatigue curve ($p = 10^2$).

5.11 Probabilistic approach for creep-fatigue model of composites

5.11.1 Main assumptions

Traditional computational approaches are deterministic in nature; they do not account for uncertainties associated with composite structures and materials. It appears inevitable that the structural engineering community, as well as many other engineering communities that are ultimately responsible for life safety issues, will eventually incorporate probabilistic analysis methods to some degree. Probabilistic analysis methods, unlike traditional deterministic methods, provide a means to quantify the inherent risk of a structural design and to quantify the sensitivities of the most important parts of the design in the overall reliability of the structural system as a whole. The degree to which these methods are successfully applied depends on addressing the issues and concerns discussed in this book. Certainly, one issue is to disseminate familiarity and the basic understanding of the principals and assumptions made in probability-based structural design. Probability—based structural engineering design, relies on applied probability methods to assess reliability within a given design criterion (for instance, by providing a limiting value to the so-called reliability index β). This allows, in turn, provision of the risk assessment to be included into structural design process. In this regard, probabilistic design methods are essential for modern engineering that enables solving complex engineering problems in the face of uncertainty. The importance of the probabilistic approach in the issues of the cyclic creep-fatigue process and the long-term behavior with a specified reliability is addressed in NASA report [11], technological solutions which will ensure the economic ability and environmental compatibility of a future High

Speed Civil Transport (HSCT) plane are currently being sought. Lighter structural concepts for both airframe primary structures and engine structure components are being investigated. It is envisioned that such objectives can be achieved through the use of composites as well as other conventional lighter weight alloys. One of the prime issues for these structural components is assured long-term behavior with a specified reliability. For example, the engine structure components are required to last 18 000 hrs, and the airframe structural components are required to last 60 000 hrs of flight time. These objectives are being pursued with the cooperation of industry and university participants through developments in accelerated testing strategies as well as computational simulation methodologies for predicting the long term durability of HSCT structural components. The present chapter addresses issues pertaining to the probabilistic fatigue life of composites under combined thermal/mechanical cyclic loads through some typical examples. Traditional computational approaches are deterministic in nature; they do not account for uncertainties associated with composite structures and materials. The focus of ongoing research at the NASA Glenn Research Center has been to develop advanced integrated computational methods and related computer codes to perform a complete probabilistic assessment of composite structures. These methods account for uncertainties in all the constituent properties, fabrication process variables, and loads to predict probabilistic micro, ply – laminate, and structural responses. These methods have already been implemented in the Integrated Probabilistic Assessment of Composite Structures (IPACS) computer code [18]. The deterministic simulations of these models are validated experimentally with fatigue data of composites or nanocomposites. Then, the models are treated probabilistically by considering the models' parameters in stochastic forms. Usually the Monte Carlo (MC) simulation is employed to investigate the models under stochastic conditions. It requires a prior knowledge of probabilistic distributions of these parameters and its dependence (or independence) with each other. It makes the practical application of the (MC) method very difficult and expensive. This chapter quantifies the cyclic creep-fatigue propagation process by creating a subset of random functions (assuming that the parameter α_i in Equation (5.1) is a random variable). For a detailed explanation on this subject please see the authors work [19]. The proposed procedure is useful for selecting a proper probabilistic fatigue model in specific applications and can be used in future fatigue studies not only in the automotive industry but also in other critical fields, to obtain more reliable conclusions. The prediction of fatigue life of structures is of vital importance to aircraft propulsion systems and their mission capability.

5.11.2 *Continuum damage and temperature effects on the probabilistic approach*

Continuum damage mechanics [20–26] provides an effective approach for characterizing the fatigue damage evolution and enables prediction of the life of composite components due to various high temperature effects on the durability of structural elements. The fatigue model developed in this book for predicting the fatigue life of composites and nanocomposites incorporates applied maximum stress, stress amplitude, loading, frequency, tensile modulus, and material constants as high temperature effects on the fatigue life of composites. It can be noted that not only the amplitude of stresses a_0 can be considered as an independent random variable, but the damage function ω (θ) can also be considered as a random function. In this case, the limit state of fatigue of the composite (or nanocomposite) is the condition ω (θ) = 1. It should also be noted that both of these functions can be considered as ergodic functions of the dimensionless temperature. In this case, the first random fatigue function (S – N) has a larger autocovariance value, and the second one has a small covariance value (or standard variance in the case of a normal probability distribution). From all of the above, it follows that the calculation of the fatigue probability according to the first scheme leads to less reliable results than according to the second scheme ω = 1. To confirm these considerations, the corresponding examples are given below. Considering, as before (see[3]) that a discrete random variable is the dimensionless parameter of the relaxation time αi from the interval (0.001 – 100000), obtain a family of fatigue curves and, analytical expressions of ω (θ) and a0 (Nf).

5.12 Examples

Example 5.11—α = 0.001

Calculated values of DEQ variables

	Variable	Initial value	Minimal value	Maximal value	Final value
1	A	2.	2.	2.	2.
2	A0	2.	2.	2.	2.
3	A1	1.	1.	1.	1.
4	A2	0	0	0.0486693	0.0486693
5	A3	0	0	0.5633915	0.5633915
6	c1	3.	3.	3.	3.

7	c2	5.	5.	5.	5.
8	d	0	0	7.406734	7.406734
9	deff	0	0	0.1107758	0.1107758
10	deff1	0	0	0.5633915	0.5633915
11	E	1.45	0.3625003	1.45	0.3625003
12	f3	1.	1.	1.	1.
13	f31	0	0	0.9999711	0.274635
14	f32	1.	0.7435871	1.	0.7435871
15	f33	0	0	0.9999998	0.2042174
16	f34	0	0	0.857183	0.857183
17	f35	1.	0	1.	0
18	f36	0	0	1.	1.
19	fi	0.05	0.05	0.05	0.05
20	k	10.	10.	10.	10.
21	m	0	0	0.0740673	**0.0740673**
22	m1	0.01	0.01	0.01	0.01
23	m2	0	0	0.0825484	0.0802933
24	m20	0	–4.606774	0.0074921	–4.606774
25	m201	0.0538	–1.270093	0.0538	–1.270093
26	m21	0.0868	0.0007299	0.0868	0.0038857
27	n	1.	1.	1.	1.
28	p	100.	100.	100.	100.
29	s	1.	1.	1.	1.
30	t	0	0	7.5	7.5
31	t1	4.	4.	4.	4.
32	t2	6.	6.	6.	6.
33	X	0	0	0.9914823	**0.9914823**
34	X1	0	0	0.0512215	0.0512215
35	X2	0	0	0.9402608	0.9402608
36	Y1	0	–0.207308	0.3624689	0.0817113
37	z	0.175	0.0598948	291.2153	291.2153
38	z1	0	–0.7197266	0	–0.7197266
39	z11	0	–0.0805399	5.895E+05	5.895E+05
40	z12	0	–0.0871341	2.828E+05	2.828E+05

41	z2	0	–0.0032159	2.124E-05	-0.0032159
42	z21	0	0	6.597E+05	6.597E+05
43	z22	0	–8.672E-05	3.164E+05	3.164E+05
44	z3	0	–290.259	0	–290.259
45	z31	0	–0.1133602	9.714E+08	9.714E+08
46	z32	0	–0.4821961	4.503E+08	4.503E+08

Differential equations

1 d(z11)/d(t) = (1/(1-X)^6)*(exp(t/(1+0.067*t)))*A1*((exp(0.001*m)))*m1*(sin (p*m))*z^n

2 d(z12)/d(t) = (1/(1-X)^6)*(exp(t/(1+0.067*t)))*A1*((exp(0.001*m)))*m1*(cos (p*m))*z^n

3 d(z21)/d(t) = (1/(1-X)^6)*(1/k)*(exp(t/(1+0.067*t)))*(A2)*(z^s)*((exp(0.001*m)))*m1*(sin (p*m))*z^n

4 d(z22)/d(t) = (1/(1-X)^6)*(1/k)*(exp(t/(1+0.067*t)))*(A2)*(z^s)*((exp(0.001*m)))*m1*(cos (p*m))*z^n

5 d(z31)/d(t) = (1/(1-X)^6)*(1/k)*(exp(t/(1+0.067*t)))*A3*(z^s)*f3*((exp(0.001*m20)))*m201*((sin (p*m20))^1)*z^n

6 d(z32)/d(t) = (1/(1-X)^6)*(1/k)*(exp(t/(1+0.067*t)))*A3*f3*(z^s)*((exp(0.001*m20)))*m201*1*(cos (p*m20))*z^n

7 d(X1)/d(t) = m1*A*(z^2)/((1-X1)^6)

8 d(X2)/d(t) = m20*A0*(z^2)/((1-X2)^2)

Explicit equations

1 m = 0.01*t

2 m1 = 0.01

3 m20 = 0.0538*t - 0.095*(t^2) + 0.000508*(t^3)

4 p = (10^2)

5 fi = 0.05

6 c1 = 3*(1-exp(-0.4*t))^0

7 c2 = 5*(1-exp(-0.4*t))^0

8 t2 = 6

9 t1 = 4

10 k = 10

11 n = 1.0

12 s = 1.0

```
13 E = (0.625-0.375*(tanh(c1*(t-t1))))*(1-fi)+(k)*(0.625-0.375*
   (tanh(c2*(t-t2))))*(fi)
14 z1 = (cos(p*m))*z11-(sin(p*m))*z12
15 z2 = (cos(p*m))*z21-(sin(p*m))*z22
16 Y1 = (0.25)*(sin (p*m))*E
17 A1 = 1
18 deffl = (1+fi*((10*t+1)^0.333-1))^3 - 1
19 d = 1*t
20 A2 = 0.1*(exp(-0.04*t))*((1+2.5*fi+14.7*(fi^2)))*deffl^1
21 deff = (fi*d)/(1 + 0.333*(1 - fi)*d)
22 z3 = (cos(p*m20))*z31-(sin(p*m20))*z32
23 X = X1+X2
24 A = 2
25 A0 = 2
26 f3 = if (t<0) then (0) else (1)
27 f31 = (1/16)*t*(8-t)
28 f32 = (exp(-0.04*t))
29 f33 = (sin(3.14*t/4))^2
30 f34 = (1/64)*t^2
31 f35 = if (t<4) then (1) else (0)
32 f36 = if (t<4) then (0) else (1)
33 m2 = 0.0868*t - 0.0386*(t^2)+ 0.0080*(t^3) - 0.000756*(t^4) +
   2.6*(10^-5)*(t^5)
34 m21 = 0.0868 - 0.0772*t + 0.024*(t^2) - 0.003*(t^3) + 1.3*(10^-4)*
   (t^4)
35 A3 = deffl
36 z = Y1-(z1+z2)*((exp(-0.001*m)))-z3*(exp(-0.001*m20))+0.175*1
37 m201 = 0.0538- 0.19*t + 0.00152*(t^2)
```

Model: X = a1*t + a2*t^2 + a3*t^3 + a4*t^4

Variable	Value
a1	–0.0159137
a2	0.0401347
a3	–0.0107286
a4	0.0008204

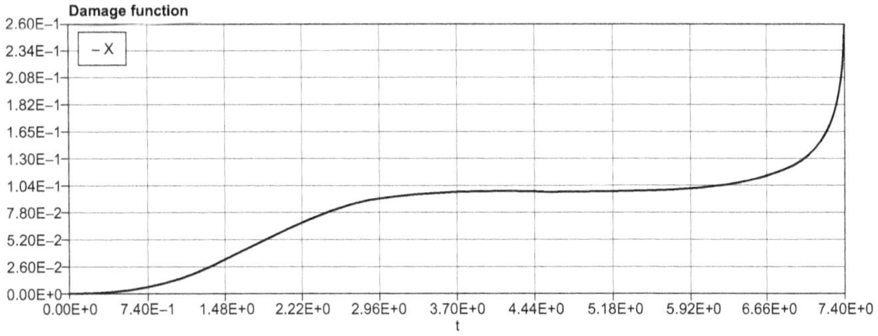

Figure 5.3. Damage function; $\alpha = 0.001$.

$$\omega = X = -\,0.0158\,\theta + 0.0401\theta^2 - 0.0107\,\theta^3 + 0.00082\theta^4 \tag{5.1}$$

$\alpha = 0.01$

Differential equations

1 d(z11)/d(t) = (1/(1-X)^6)*(exp(t/(1+0.067*t)))*A1*((exp(0.01*m)))*m1 *(sin (p*m))*z^n

2 d(z12)/d(t) = (1/(1-X)^6)*(exp(t/(1+0.067*t)))*A1*((exp(0.01*m)))*m 1*(cos (p*m))*z^n

3 d(z21)/d(t) = (1/(1-X)^6)*(1/k)*(exp(t/(1+0.067*t)))*(A2)*(z^s)* ((exp(0.01*m)))*m1*(sin (p*m))*z^n

4 d(z22)/d(t) = (1/(1-X)^6)*(1/k)*(exp(t/(1+0.067*t)))*(A2)*(z^s)* ((exp(0.01*m)))*m1*(cos (p*m))*z^n

5 d(z31)/d(t) = (1/(1-X)^6)*(1/k)*(exp(t/(1+0.067*t)))*A3*(z^s)*f3* ((exp(0.01*m20)))*m201*((sin (p*m20))^1)*z^n

6 d(z32)/d(t) = (1/(1-X)^6)*(1/k)*(exp(t/(1+0.067*t)))*A3*f3*(z^s)* ((exp(0.01*m20)))*m201*1*(cos (p*m20))*z^n

7 d(X1)/d(t) = m1*A*(z^2)/((1-X1)^6)

8 d(X2)/d(t) = m20*A0*(z^2)/((1-X2)^2)

Model: X = a1*t + a2*t^2 + a3*t^3 + a4*t^4

Variable	Value
a1	–0.0132798
a2	0.0376365
a3	–0.0100519
a4	0.0007654

$$\boldsymbol{\omega} = X = -0.0133\,\theta + 0.0376\,\theta^2 - 0.0100\theta^3 + 0.00076\theta^4 \quad (5.2)$$

$\alpha = \mathbf{1.0}$

Differential equations

1 d(z11)/d(t) = (1/(1-X)^6)*(exp(t/(1+0.067*t)))*A1*((exp(1*m)))*m1*(sin (p*m))*z^n

2 d(z12)/d(t) = (1/(1-X)^6)*(exp(t/(1+0.067*t)))*A1*((exp(1*m)))*m1*(cos (p*m))*z^n

3 d(z21)/d(t) = (1/(1-X)^6)*(1/k)*(exp(t/(1+0.067*t)))*(A2)*(z^s)* ((exp(1*m)))*m1*(sin (p*m))*z^n

4 d(z22)/d(t) = (1/(1-X)^6)*(1/k)*(exp(t/(1+0.067*t)))*(A2)*(z^s)* ((exp(1*m)))*m1*(cos (p*m))*z^n

5 d(z31)/d(t) = (1/(1-X)^6)*(1/k)*(exp(t/(1+0.067*t)))*A3*(z^s)*f3* ((exp(1*m20)))*m201*((sin (p*m20))^1)*z^n

6 d(z32)/d(t) = (1/(1-X)^6)*(1/k)*(exp(t/(1+0.067*t)))*A3*f3*(z^s)* ((exp(1*m20)))*m201*1*(cos (p*m20))*z^n

7 d(X1)/d(t) = m1*A*(z^2)/((1-X1)^6)

8 d(X2)/d(t) = m20*A0*(z^2)/((1-X2)^2)

Model: X = a1*t + a2*t^2 + a3*t^3 + a4*t^4

Variable	Value
a1	–0.0260985
a2	0.0489586
a3	–0.0129066
a4	0.0009804

$$\boldsymbol{\omega} = X = -0.0261[\theta] + 0.0490\theta^2 - 0.0129\,\theta^3 + 0.00098\theta^4 \quad (5.3)$$

$\alpha = 10$

Differential equations

1 d(z11)/d(t) = (1/(1-X)^6)*(exp(t/(1+0.067*t)))*A1*((exp(10*m)))*m1* (sin (p*m))*z^n

2 d(z12)/d(t) = (1/(1-X)^6)*(exp(t/(1+0.067*t)))*A1*((exp(10*m)))*m1* (cos (p*m))*z^n

3 d(z21)/d(t) = (1/(1-X)^6)*(1/k)*(exp(t/(1+0.067*t)))*(A2)*(z^s)* ((exp(10*m)))*m1*(sin (p*m))*z^n

4 d(z22)/d(t) = (1/(1-X)^6)*(1/k)*(exp(t/(1+0.067*t)))*(A2)*(z^s)*((exp(10*m)))*m1*(cos (p*m))*z^n

5 d(z31)/d(t) = (1/(1-X)^6)*(1/k)*(exp(t/(1+0.067*t)))*A3*(z^s)*f3*((exp(10*m20)))*m201*((sin (p*m20))^1)*z^n

6 d(z32)/d(t) = (1/(1-X)^6)*(1/k)*(exp(t/(1+0.067*t)))*A3*f3*(z^s)*((exp(10*m20)))*m201*1*(cos (p*m20))*z^n

7 d(X1)/d(t) = m1*A*(z^2)/((1-X1)^6)

8 d(X2)/d(t) = m20*A0*(z^2)/((1-X2)^2)

Model: X = a1*t + a2*t^2 + a3*t^3 + a4*t^4

Variable	Value
a1	–0.0286921
a2	0.0509674
a3	–0.013363
a4	0.0010095

$$\omega = X = -\,0.0287\theta + 0.0510\,\theta^2 - 0.0134\theta^3 + 0.0010\theta^4 \qquad (5.4)$$

$\alpha = 100$

Differential equations

1 d(z11)/d(t) = (1/(1-X)^6)*(exp(t/(1+0.067*t)))*A1*((exp(100*m)))*m1*(sin (p*m))*z^n

2 d(z12)/d(t) = (1/(1-X)^6)*(exp(t/(1+0.067*t)))*A1*((exp(100*m)))*m1*(cos (p*m))*z^n

3 d(z21)/d(t) = (1/(1-X)^6)*(1/k)*(exp(t/(1+0.067*t)))*(A2)*(z^s)*((exp(100*m)))*m1*(sin (p*m))*z^n

4 d(z22)/d(t) = (1/(1-X)^6)*(1/k)*(exp(t/(1+0.067*t)))*(A2)*(z^s)*((exp(100*m)))*m1*(cos (p*m))*z^n

5 d(z31)/d(t) = (1/(1-X)^6)*(1/k)*(exp(t/(1+0.067*t)))*A3*(z^s)*f3*((exp(100*m20)))*m201*((sin (p*m20))^1)*z^n

6 d(z32)/d(t) = (1/(1-X)^6)*(1/k)*(exp(t/(1+0.067*t)))*A3*f3*(z^s)*((exp(100*m20)))*m201*1*(cos (p*m20))*z^n

7 d(X1)/d(t) = m1*A*(z^2)/((1-X1)^6)

8 d(X2)/d(t) = m20*A0*(z^2)/((1-X2)^2)

Model: X = a1*t + a2*t^2 + a3*t^3 + a4*t^4

Variable	Value
a1	–0.0153552
a2	0.0370251
a3	–0.0094526
a4	0.0006802

$$\boldsymbol{\omega} = X = -0.0154\boldsymbol{\theta} + 0.0370\boldsymbol{\theta}^2 - 0.0094\boldsymbol{\theta}^3 + 0.00068\boldsymbol{\theta}^4 \quad (5.5)$$

$\alpha = 1000$

Differential equations

1 d(z11)/d(t) = (1/(1-X)^6)*(exp(t/(1+0.067*t)))*A1*((exp(1000*m)))*m 1*(sin (p*m))*z^n

2 d(z12)/d(t) = (1/(1-X)^6)*(exp(t/(1+0.067*t)))*A1*((exp(1000*m)))*m 1*(cos (p*m))*z^n

3 d(z21)/d(t) = (1/(1-X)^6)*(1/k)*(exp(t/(1+0.067*t)))*(A2)*(z^s)* ((exp(1000*m)))*m1*(sin (p*m))*z^n

4 d(z22)/d(t) = (1/(1-X)^6)*(1/k)*(exp(t/(1+0.067*t)))*(A2)*(z^s)* ((exp(1000*m)))*m1*(cos (p*m))*z^n

5 d(z31)/d(t) = (1/(1-X)^6)*(1/k)*(exp(t/(1+0.067*t)))*A3*(z^s)*f3* ((exp(1000*m20)))*m201*((sin (p*m20))^1)*z^n

6 d(z32)/d(t) = (1/(1-X)^6)*(1/k)*(exp(t/(1+0.067*t)))*A3*f3*(z^s)* ((exp(1000*m20)))*m201*1*(cos (p*m20))*z^n

7 d(X1)/d(t) = m1*A*(z^2)/((1-X1)^6)

8 d(X2)/d(t) = m20*A0*(z^2)/((1-X2)^2)

Model: X = a1*t + a2*t^2 + a3*t^3 + a4*t^4

Variable	Value
	0.0119006
a2	0.006964
a3	–0.0013808
a4	6.694E-05

$$\boldsymbol{\omega} = X = 0.0119\theta + 0.00696\,\theta^2 - 0.00138\theta^3 + 0.000067\,\theta^4 \quad (5.6)$$

$\alpha = 10000$

Differential equations

1 d(z11)/d(t) = (1/(1-X)^6)*(exp(t/(1+0.067*t)))*A1*((exp(10000*m)))*m1*(sin (p*m))*z^n

2 d(z12)/d(t) = (1/(1-X)^6)*(exp(t/(1+0.067*t)))*A1*((exp(10000*m)))*m1*(cos (p*m))*z^n

3 d(z21)/d(t) = (1/(1-X)^6)*(1/k)*(exp(t/(1+0.067*t)))*(A2)*(z^s)*((exp(10000*m)))*m1*(sin (p*m))*z^n

4 d(z22)/d(t) = (1/(1-X)^6)*(1/k)*(exp(t/(1+0.067*t)))*(A2)*(z^s)*((exp(10000*m)))*m1*(cos (p*m))*z^n

5 d(z31)/d(t) = (1/(1-X)^6)*(1/k)*(exp(t/(1+0.067*t)))*A3*(z^s)*f3*((exp(10000*m20)))*m201*((sin (p*m20))^1)*z^n

6 d(z32)/d(t) = (1/(1-X)^6)*(1/k)*(exp(t/(1+0.067*t)))*A3*f3*(z^s)*((exp(10000*m20)))*m201*1*(cos (p*m20))*z^n

7 d(X1)/d(t) = m1*A*(z^2)/((1-X1)^6)

8 d(X2)/d(t) = m20*A0*(z^2)/((1-X2)^2)

Model: X = a1*t + a2*t^2 + a3*t^3 + a4*t^4

Variable	Value
a1	–0.0013819
a2	0.0257224
a3	–0.0069785
a4	0.0005197

$$\boldsymbol{\omega} = X = -\,0.0014\,\theta + 0.025\,\theta^{2} - 0.00698\,\theta^{3} + 0.000520\,\theta^{4} \qquad (5.7)$$

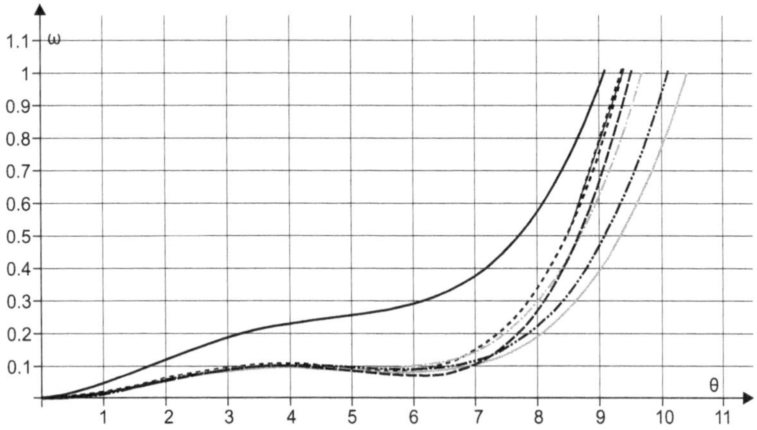

Figure 5.4. Damage function (Summary: *ergodic* random process) α = 1000.

One can see that the random process ω (θ) can be considered as **ergodic in the wide sense**, therefore only one realization is enough to compute mean μ_ω and standard deviation σ_ω parameters (normal distribution assumed here) for the so-called correlation theory application. The computations are as follows:

$$\boldsymbol{\omega} = X = -0.0287*t + 0.0510*t\text{^}2 - 0.0134*t\text{^}3 + 0.0010*t\text{^}4 \quad (5.8)$$

$$\boldsymbol{\omega} = X = -0.0261\,\theta + 0.0490\theta^2 - 0.0129\,\theta^3 + 0.00098\theta^4 \quad (5.3)$$

$a_0 = 0.25$; $p = 10^2$;

$$\mu_\omega = \frac{1}{\theta_{max}} \int_0^{\theta\,max} \omega(\tau)d\tau = \frac{1}{9}\int_0^9 (\mathbf{-0.0287\theta + 0.051\theta^2 - 0.0134\theta^3 + 0.001\theta^4})\mathbf{d\theta = 0.118}$$

$$\sigma_\omega^2 = \frac{1}{\theta_{max}} \int_0^{\theta max} [\omega(\tau) - \mu_\omega]^2 d\tau =$$

$$= \frac{1}{9}\int_0^9 ((\mathbf{-0.0287\theta} + 0.051\boldsymbol{\theta^2} - 0.0134\boldsymbol{\theta^3} + \mathbf{0.001\theta^4}) - 0.118)^2 d\theta = \mathbf{0.01}7$$

$$\sigma_\omega = \sqrt{\mathbf{0.01}7} = 0.13; \quad \boldsymbol{\beta} = \frac{1 - 0.118}{0.13} = 6.78$$

In order to calculate the standard deviation in this case the chosen function has to be centered.

Assuming that composite (or nanocomposite) material fails if **ω** = 1, one can compute the probability of failure (or reliability $P_r = 1 - P_f$).

Example 5.13—p = 10000; random process **ω** (a_0) is an ergodic process: $\alpha = 10$

$$X = 0.212a_0 + 12.16(a_0)^2 - 20.53(a_0)^3 + 12.32(a_0)^4 \quad (5.8)$$

$$\boldsymbol{\omega} = X = -0.0261\,\theta + 0.0490\theta^2 - 0.0129\,\theta^3 + 0.00098\theta^4$$

$$\mu_{a0} = 1/0.8 \int_{0.15}^{0.95} [\mathbf{0.212a_0 - 2.76(a_0) + 12.16(a_0)^2 - 20.53(a_0)^3 + 12.32(a_0)^4}]\mathbf{da_0} = 0.06898/0.8 = 0.0862 \quad (5.9)$$

$$\sigma_{a0}^2 = 1/0.8$$

$$\int_{0.15}^{0.95} [\mathbf{0.212a_0 - 2.76(a_0) + 12.16(a_0)^2 - 20.53(a_0)^3 + 12.32(a_0)^4 - 0.0862}]^2 da_0$$

$$= 0.07330 \quad (5.10)$$

$\sigma_{a0} = 0.271$

$P_{fa0} = 1 - \Phi^*((1 - 0.0862)/0.271) = 1 - 0.99965 = \mathbf{3.5(10^{-4})}$

The FAA has the "Acceptable Probability of Failure" (per FAA Database).

Table 5.2. Acceptable probability of failure (per FAA Database).

Type	Number	Probability of Failures (per Flight)
Total Incidents	5,497	$5.5(10^{-5})$
Structural Incidents	62	$6.2(10^{-7})$
Structural Accidents	1	$1(10^{-8})$

If the reliability index β is known or given (see Table 5.1) above then $\Phi^*(-\beta) = P_f$ can be easily computed (see Table below).

Table 5.3. Reliability index β vs. probability of failure P_f.

$\Phi^*(-\beta)$	0.1	0.01	$1(10^{-3})$	$1(10^{-4})$	$1(10^{-5})$	$1(10^{-6})$	$1(10^{-7})$
β	1.28	2.32	3.15	3.77	4.0	4.5	5.0

If $\boldsymbol{\beta} > 5$: $P_f = \boldsymbol{\Phi}^*(-\boldsymbol{\beta}) = [1/\sqrt{2\boldsymbol{\pi}}](\boldsymbol{\beta}^2 - 1)/\boldsymbol{\beta}^3\}\exp(-\boldsymbol{\beta}^2/2)$ (5.11)

$\boldsymbol{\beta} = (1 - 0.118)/0.13 = \mathbf{6.78} > 5$ and $P_f = \Phi^*(-6.78) = 0.4\{(6.78^2 - 1)/6.78^3\} \exp(-6.78^2/2) = 6.0(10^{-12}) < < (10^{-8})$ Fatigue limit (number of cycles) $N_f = (pm)/2\pi = 0.08(100)/2\pi = 1.27$. Maximum temperature at fatigue failure: $T_{max} = \theta_{max}(0.1T_*) + T_* = 9(0.1)600 + 600 = 1140[^0K] \approx 840\ [^0C]$

Consider now the case that the inverse probability $P_f = 10^{-8}$ is given, then:

$P_f = [1/\sqrt{2\boldsymbol{\pi}}]\ \{(\boldsymbol{\beta}^2 - 1)/\boldsymbol{\beta}^3\}\exp(-\boldsymbol{\beta}^2/2) = 10^{-8}$ (5.12)

Solving Equation (5.12) we have: $\boldsymbol{\beta} \approx \mathbf{5.7}$

$\ln[1/\sqrt{2\boldsymbol{\pi}}] + \ln\{(\boldsymbol{\beta}^2 - 1)/\boldsymbol{\beta}^3\} - (\boldsymbol{\beta}^2/2) = -18.4$ (5.13)

One can see from "forward" solution that $\beta >> 1$, therefore the Equation (5.11) can be approximately reduced to: $\ln\{1/\boldsymbol{\beta}\} - (\boldsymbol{\beta}^2/2) = -18.8$. Let's try $\beta = 5.7 \rightarrow -18 \approx -18.4$. Now the final result is: maximum dimensionless creep-fatigue temperature $\boldsymbol{\theta}_f = 8 - \boldsymbol{\beta}\,\sigma_\omega = 8 - 5.7(0.0507) = 7.71 < 8.0 \rightarrow N_f = 7.71(0.08)(10^2)/(2\boldsymbol{\pi}) = 9.82$. Maximum temperature: $T = 7.71(60) + 600 = 1062.6^0K \approx 762^0C$. Similarly, analytical models can be computed in cases of Weibull and lognormal distributions of the fatigue limit, which allow predicting cyclic durability and constructing probability fatigue curves. The construction of such models based on a joint analysis of reference data on the characteristics of fatigue resistance and the results of full-scale fatigue tests of prototypes or their analogues makes it possible to carry out resource design of the critical elements of aerospace structures.

Example 5.13—p = 10000

Differential equations

1 d(z11)/d(t) = (1/(1-X)^6)*(exp(t/(1+0.067*t)))*A1*((exp(0.1*m)))*m1*(sin (p*m))*z^n

2 d(z12)/d(t) = (1/(1-X)^6)*(exp(t/(1+0.067*t)))*A1*((exp(0.1*m)))*m1*(cos (p*m))*z^n

3 d(z21)/d(t) = (1/(1-X)^6)*(exp(t/(1+0.067*t)))*(A2)*(z^s)*((exp(0.1*m)))*m1*(sin (p*m))*z^n

4 d(z22)/d(t) = (1/(1-X)^6)*(exp(t/(1+0.067*t)))*(A2)*(z^s)*((exp(0.1*m)))*m1*(cos (p*m))*z^n

5 d(z31)/d(t) = (1/(1-X)^6)*(1/k)*(exp(t/(1+0.067*t)))*A3*(z^s)*f3*((exp(0.1*m2)))*m21*((sin (p*m2))^1)*z^n

6 d(z32)/d(t) = (1/(1-X)^6)*(1/k)*(exp(t/(1+0.067*t)))*A3*f3*(z^s)*((exp(0.1*m2)))*m21*1*(cos (p*m2))*z^n

7 d(X1)/d(t) = m*A*(z^2)/((1-X1)^6)

8 d(X2)/d(t) = m2*A0*(z^2)/((1-X2)^2)

Model: X = a1*t + a2*t^2 + a3*t^3 + a4*t^4

Variable	Value
a1	–0.1245097
a2	0.2067218
a3	–0.0783376
a4	0.0092241

$$X = \omega = -0.124\theta + 0.207\theta^2 - 0.0783\,\theta^3 + 9.22(10^{-3})\theta^4 \qquad (5.20)$$

$N_f = 7.97 = m_{max}p/2\pi = 0.501(100)/2\pi = 7.97$ (it can be also counted from Fig. 4.1) and $T_f = 912[^0K]$ ($T_f = \beta T_* \theta_{max} + 600$; $\beta T_* = 0.1\,(600)$); $\theta_{max} = 5.2$

Example 5.14—same as Example 5.1, but σ_{st} is 0 and $a_0 = 0.45$

Differential equations

1 d(z11)/d(t) = (1/(1-X)^6)*(exp(t/(1+0.067*t)))*A1*((exp(0.1*m)))*m1*(sin (p*m))*z^n

2 d(z12)/d(t) = (1/(1-X)^6)*(exp(t/(1+0.067*t)))*A1*((exp(0.1*m)))*m1*(cos (p*m))*z^n

3 d(z21)/d(t) = (1/(1-X)^6)*(exp(t/(1+0.067*t)))*(A2)*(z^s)*((exp(0.1*m)))*m1*(sin (p*m))*z^n

4 d(z22)/d(t) = (1/(1-X)^6)*(exp(t/(1+0.067*t)))*(A2)*(z^s)*((exp(0.1*m)))*m1*(cos (p*m))*z^n

5 d(z31)/d(t) = (1/(1-X)^6)*(1/k)*(exp(t/(1+0.067*t)))*A3*(z^s)*f3*((exp(0.1*m2)))*m21*((sin (p*m2))^1)*z^n

6 d(z32)/d(t) = (1/(1-X)^6)*(1/k)*(exp(t/(1+0.067*t)))*A3*f3*(z^s)*((exp(0.1*m2)))*m21*1*(cos (p*m2))*z^n

7 d(X1)/d(t) = m*A*(z^2)/((1-X1)^6)

8 d(X2)/d(t) = m2*A0*(z^2)/((1-X2)^2)

Model: X = a1*t + a2*t^2 + a3*t^3 + a4*t^4

Variable	Value
a1	–0.1515534
a2	0.3409575
a3	–0.1709551
a4	0.0281394

$$X = \omega = -\,0.152\theta + 0.341\theta^2 - 0.171\theta^3 + 2.82(10^{-2})\theta^4 \tag{5.21}$$

$N_f = 5.62 = m_{max}p/2\pi = 0.353(100)/2\pi = 5.62$ (it can be also counted from Fig. 4.1) and $T_f = 816[^0K]$ ($T_f = \beta T_* \theta_{max} + 600$; $\beta T_* = 0.1\ (600)$); $\theta_{max} = 3.6$

Example 5.15—same as Example 5.1, but σ_{st} is 0 and $a_0 = 0.55$

Differential equations

1 d(z11)/d(t) = (1/(1-X)^6)*(exp(t/(1+0.067*t)))*A1*((exp(0.1*m)))*m1*(sin (p*m))*z^n

2 d(z12)/d(t) = (1/(1-X)^6)*(exp(t/(1+0.067*t)))*A1*((exp(0.1*m)))*m1*(cos (p*m))*z^n

3 d(z21)/d(t) = (1/(1-X)^6)*(exp(t/(1+0.067*t)))*(A2)*(z^s)*((exp(0.1*m)))*m1*(sin (p*m))*z^n

4 d(z22)/d(t) = (1/(1-X)^6)*(exp(t/(1+0.067*t)))*(A2)*(z^s)*((exp(0.1*m)))*m1*(cos (p*m))*z^n

5 d(z31)/d(t) = (1/(1-X)^6)*(1/k)*(exp(t/(1+0.067*t)))*A3*(z^s)*f3*((exp(0.1*m2)))*m21*((sin (p*m2))^1)*z^n

6 d(z32)/d(t) = (1/(1-X)^6)*(1/k)*(exp(t/(1+0.067*t)))*A3*f3*(z^s)*((exp(0.1*m2)))*m21*1*(cos (p*m2))*z^n

7 d(X1)/d(t) = m*A*(z^2)/((1-X1)^6)

8 d(X2)/d(t) = m2*A0*(z^2)/((1-X2)^2)

Model: X = a1*t + a2*t^2 + a3*t^3 + a4*t^4

Variable	Value
a1	–0.2270152
a2	0.7673331
a3	–0.6396274
a4	0.1756898

$$X = \omega = -\,0.227\theta + 0.767\theta^2 - 0.639\theta^3 + 1.76(10^{-1})\,\theta^4 \qquad (5.22)$$

$N_f = 3.565 = m_{max}p/2\pi = 0.224(100)/2\pi = 3.565$ (it can be also counted from Fig. 4.1) and $T_f = 738[^0K]$

$(T_f) = \beta T_* \,\theta_{max} + 600$; $\beta T_* = 0.1\ (600))$; $\theta_{max} = 2.3$

Example 5.15—same as Example 5.1, but σ_{st} is 0 and $a_0 = 0.75$

Differential equations

1 d(z11)/d(t) = (1/(1-X)^6)*(exp(t/(1+0.067*t)))*A1*((exp(0.1*m)))*m1*(sin (p*m))*z^n

2 d(z12)/d(t) = (1/(1-X)^6)*(exp(t/(1+0.067*t)))*A1*((exp(0.1*m)))*m1*(cos (p*m))*z^n

3 d(z21)/d(t) = (1/(1-X)^6)*(exp(t/(1+0.067*t)))*(A2)*(z^s)*((exp(0.1*m)))*m1*(sin (p*m))*z^n

4 d(z22)/d(t) = (1/(1-X)^6)*(exp(t/(1+0.067*t)))*(A2)*(z^s)*((exp(0.1*m)))*m1*(cos (p*m))*z^n

5 d(z31)/d(t) = (1/(1-X)^6)*(1/k)*(exp(t/(1+0.067*t)))*A3*(z^s)*f3*((exp(0.1*m2)))*m21*((sin (p*m2))^1)*z^n

6 d(z32)/d(t) = (1/(1-X)^6)*(1/k)*(exp(t/(1+0.067*t)))*A3*f3*(z^s)*((exp(0.1*m2)))*m21*1*(cos (p*m2))*z^n

7 d(X1)/d(t) = m*A*(z^2)/((1-X1)^6)

8 d(X2)/d(t) = m2*A0*(z^2)/((1-X2)^2)

Model: X = a1*t + a2*t^2 + a3*t^3 + a4*t^4

Variable	Value
a1	–0.3427825
a2	1.741289
a3	–2.226269
a4	0.9210042

$$X = \omega = -\,0.342\theta + 1.741\theta^2 - 2.226\theta^3 + 9.21(10^{-1})\,\theta^4 \qquad (5.23)$$

$N_f = 2.387 = m_{max}p/2\pi = 0.15(100)/2\pi = 2.387$ (it can be also counted from Fig. 4.1) and $T_f = 690[^0K]$ ($T_f = \beta T_*\,\theta_{max} + 600$; $\beta T_* = 0.1\ (600)$); $\theta_{max} = 1.5$

Example 5.16—same as Example 5.1, but σ_{st} is 0 and $\mathbf{a_0 = 0.95}$

Differential equations

1 d(z11)/d(t) = (1/(1-X)^6)*(exp(t/(1+0.067*t)))*A1*((exp(0.1*m)))*m1 *(sin (p*m))*z^n

2 d(z12)/d(t) = (1/(1-X)^6)*(exp(t/(1+0.067*t)))*A1*((exp(0.1*m)))*m1 *(cos (p*m))*z^n

3 d(z21)/d(t) = (1/(1-X)^6)*(exp(t/(1+0.067*t)))*(A2)*(z^s)* ((exp(0.1*m)))*m1*(sin (p*m))*z^n

4 d(z22)/d(t) = (1/(1-X)^6)*(exp(t/(1+0.067*t)))*(A2)*(z^s)* ((exp(0.1*m)))*m1*(cos (p*m))*z^n

5 d(z31)/d(t) = (1/(1-X)^6)*(1/k)*(exp(t/(1+0.067*t)))*A3*(z^s)*f3* ((exp(0.1*m2)))*m21*((sin (p*m2))^1)*z^n

6 d(z32)/d(t) = (1/(1-X)^6)*(1/k)*(exp(t/(1+0.067*t)))*A3*f3*(z^s)* ((exp(0.1*m2)))*m21*1*(cos (p*m2))*z^n

7 d(X1)/d(t) = m*A*(z^2)/((1-X1)^6)

8 d(X2)/d(t) = m2*A0*(z^2)/((1-X2)^2)

Model: X = a1*t + a2*t^2 + a3*t^3 + a4*t^4

Variable	Value
a1	0.0718596
a2	–0.2365555
a3	0.6840137
a4	–0.3121106

$$X = \omega = 0.0718\theta - 0.236\theta^2 + 0.684\theta^3 - 0.312(10^{-1})\theta^4 \qquad (5.24)$$

Table 5.10. Maximum dimensionless temperature θ_{max} (frequency $p = 1 - 10^7$).

a/σ_{st}	0.15	0.25	0.35	0.45	0.55	0.75	0.95
X=ω p = 1	$\omega = 1.0$ $m_{max} = 1.13$ $\theta_{max} = 11.8$ $N_f = 0.18$	$\omega = 1.0$ $m_{max} = 0.96$ $\theta_{max} = 10.0$ $N_f = 0.153$	$\omega = 1.0$ $m_{max} =$ 0.844 $\theta_{max} = 8.8$ $N_f = 0.134$	$\omega = 1.0$ $m_{max} =$ 0.756 $\theta_{max} = 7.8$ $N_f =$ 0.120	$\omega = 1.0$ $m_{max} =$ 0.682 $\theta_{max} = 7.0$ $N_f = 0.108$	$\omega = 1.0$ $m_{max} = 0.55$ $\theta_{max} = 5.5$ $N_f = 0.0875$	$\omega = 1.0$ $m_{max} =$ 0.462 $\theta_{max} = 4.8$ $N_f = 0.0735$
p = 10	$\omega = 1$ $m_{max} = 0.76$ $\theta_{max} = 8.0$ $N_f = 1.216$	$\omega = 1$ $m_{max} =$ 0.578 $\theta_{max} = 6.0$ $N_f = 0.92$	$\omega = 1$ $m_{max} =$ 0.271 $\theta_{max} = 2.80$ $N_f = 0.431$	$\omega = 1$ $m_{max} =$ 0.196 $\theta_{max} = 2.0$ $N_f = 0.312$	$\omega = 1$ $m_{max} =$ 0.166 $\theta_{max} = 1.7$ $N_f = 0.226$	$\omega = 1$ $m_{max} = 0.136$ $\theta_{max} = 1.40$ $N_f = 0.216$	$\omega = 1$ $m_{max} =$ 0.118 $\theta_{max} = 1.2$ $N_f = 0.170$
X=ω p = 10^2	$\omega = 1.0$ $m_{max} = 1.4$ $\theta_{max} = 14.0$ $N_f = 22.28$	$\omega = 1.0$ $m_{max} =$ 0.772 $\theta_{max} = 8.0$ $N_f = 12.29$	$\omega = 1.0$ $m_{max} = 0.50$ $\theta_{max} = 5.2$ $N_f = 7.97$	$\omega = 1.0$ $m_{max} =$ 0.353 $\theta_{max} = 3.2$ $N_f = 5.62$	$\omega = 1.0$ $m_{max} = 0.22$ $\theta_{max} = 2.2$ $N_f = 3.5$	$\omega = 1.0$ $m_{max} = 0.15$ $\theta_{max} = 1.5$ $N_f = 2.39$	$\omega = 1.0$ $m_{max} = 0.10$ $\theta_{max} = 1.0$ $N_f = 1.59$
X=ω p = 10^3	$\omega = 1.0$ $m_{max} = 1.66$ $\theta_{max} = 17.0$ $N_f = 264.2$	$\omega = 1.0$ $m_{max} =$ 0.835 $\theta_{max} = 8.6$ $N_f = 84.4$	$\omega = 1.0$ $m_{max} =$ 0.522 $\theta_{max} = 5.4$ $N_f = 83.09$	$\omega = 1.0$ $m_{max} =$ 0.364 $\theta_{max} = 3.7$ $N_f = 57.93$	$\omega = 1.0$ $m_{max} =$ 0.243 $\theta_{max} = 2.55$ $N_f = 38.67$	$\omega = 1.0$ $m_{max} = 0.16$ $\theta_{max} = 1.6$ $N_f = 25.5$	$\omega = 1.0$ $m_{max} =$ 0.122 $\theta_{max} = 1.25$ $N_f = 19.42$
X=ω p = 10^4	$\omega = 1$ $m_{max} = 1.65$ $\theta_{max} = 16.6$ $N_f = 26420$	$\omega = 1$ $m_{max} = 0.54$ $\theta_{max} = 5.6$ $N_f = 864.1$	$\omega = 1$ $m_{max} =$ 0.422 $\theta_{max} = 4.4$ $N_f = 671.6$	$\omega = 1$ $m_{max} = 0.3$ $\theta_{max} = 3.1$ $N_f =$ 477.5	$\omega = 1$ $m_{max} =$ 0.218 $\theta_{max} = 2.2$ $N_f = 346.9$	$\omega = 1$ $m_{max} = 0.152$ $\theta_{max} = 1.6$ $N_f = 241.9$	$\omega = 1$ $m_{max} =$ 0.121 $\theta_{max} = 1.25$ $N_f = 192.6$
X=ω p = 10^5	$\omega = 1$ $m_{max} = 16.5$ $\theta_{max} = 16.6$ $N_f = 10{,}011$	$\omega = 1$ $m_{max} = 0.54$ $\theta_{max} = 5.6$ $N_f = 8{,}641$	$\omega = 1$ $m_{max} =$ 0.422 $\theta_{max} = 4.4$ $N_f = 6{,}716$	$\omega = 1$ $m_{max} = 0.3$ $\theta_{max} = 3.1$ $N_f =$ 4,775	$\omega = 1$ $m_{max} =$ 0.218 $\theta_{max} = 2.2$ $N_f = 3{,}469$	$\omega = 1$ $m_{max} = 0.152$ $\theta_{max} = 1.6$ $N_f = 2{,}419$	$\omega = 1$ $m_{max} =$ 0.121 $\theta_{max} = 1.25$ $N_f = 1{,}926$
X=ω p = 10^6	$\omega = 1$ $m_{max} = 0.629$ $\theta_{max} = 6.6$ $N_f =$ 100,110	$\omega = 1$ $m_{max} = 0.54$ $\theta_{max} = 5.6$ $N_f =$ 86,410	$\omega = 1$ $m_{max} =$ 0.422 $\theta_{max} = 4.4$ $N_f = 67{,}160$	$\omega = 1$ $m_{max} = 0.3$ $\theta_{max} = 3.1$ $N_f =$ 47,750	$\omega = 1$ $m_{max} = 0.218$ $\theta_{max} = 2.2$ $N_f =$ 34,690	$\omega = 1$ $m_{max} = 0.152$ $\theta_{max} = 1.6$ $N_f = 24{,}190$	$\omega = 1$ $m_{max} =$ 0.121 $\theta_{max} = 1.25$ $N_f = 19{,}250$
X=ω p = 10^7	$\omega = 1$ $m_{max} = 0.629$ $\theta_{max} = 6.6$ $N_f =$ 1,001,100	$\omega = 1$ $m_{max} = 0.54$ $\theta_{max} = 5.6$ $N_f =$ 864,100	$\omega = 1$ $m_{max} = 0.422$ $\theta_{max} = 4.4$ $N_f =$ 671,600	$\omega = 1$ $m_{max} = 0.3$ $\theta_{max} = 3.1$ $N_f =$ 477,500	$\omega = 1$ $m_{max} = 0.218$ $\theta_{max} = 2.2$ $N_f =$ 346,900	$\omega = 1$ $m_{max} = 0.152$ $\theta_{max} = 1.6$ $N_f =$ 241,900	$\omega = 1$ $m_{max} = 0.121$ $\theta_{max} = 1.25$ $N_f =$ 192,500

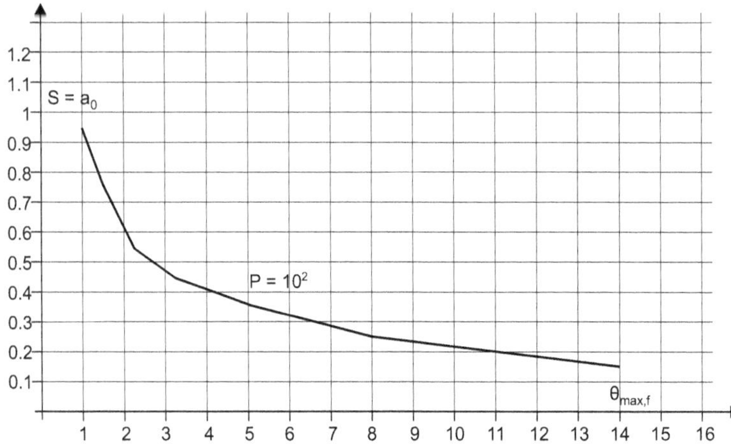

Figure 5.5. S – N fatigue curve (p = 10^2).

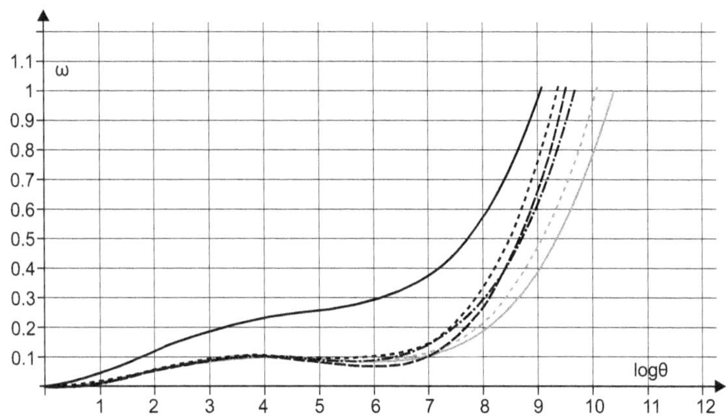

Figure 5.6. Damage Function (Summary: *ergodic* random process).

$$\boldsymbol{\omega} = X = -0.0014\,\theta + 0.025\,\theta^2 - 0.00698\,\theta^3 + 0.000520\,\theta^4 \tag{5.31}$$

$$\mu_\omega = \frac{1}{\theta_{max}} \int_0^{\theta max} \omega(\tau)d\tau = \frac{\mathbf{1}}{9} \int_0^9 \mathbf{(-0.0287\theta + 0.051\theta^2 - 0.0134\theta^3 + 0.001\theta^4)d\theta = 0.118}$$

In order to calculate the standard deviation in this case the chosen function (1) has to be centered.

Again, using the POLYMATH software the standard deviation σ is computed as follows:

$$\sigma_{\omega}^{2} = \frac{1}{\theta_{max}} \int_0^{\theta_{max}} [\omega(\tau) - \mu_{\omega}]^2 d\tau =$$

$$= \frac{1}{9} \int_0^9 ((-\mathbf{0}.0287\boldsymbol{\theta} + 0.051\boldsymbol{\theta}^2 - 0.0134\boldsymbol{\theta}^3 + \mathbf{0}.001\boldsymbol{\theta}^4) - 0.118)^2 d\theta = \mathbf{0.017}$$

$$\sigma_{\omega} = \sqrt{\mathbf{0}.017} = 0.13; \quad \beta = \frac{1 - 0.118}{0.13} = 6.78$$

Assuming that the composite (or nanocomposite) material fails if **ω** = 1, one can compute the probability of failure (or reliability $P_r = 1 - P_f$) [5.22].

Table 5.4. Acceptable Probability of Failure (per FAA Database).

Type	**Number**	**Probability of Failures (per Flight)**
Total Incidents	5,497	$5.5(10^{-5})$
Structural Incidents	62	$6.2(10^{-7})$
Structural Accidents	1	$1(10^{-8})$

The FAA has the "Acceptable Probability of Failure" (per FAA Database).

If the reliability index **β** is known or given (see Table 5.4) above then $\Phi^*(-\beta) = P_f$ can be easy computed (see Table 5.5 below)

Table 5.5. Reliability index β vs. probability of failure P_f.

$\Phi^*(-\beta)$	0.1	0.01	$1(10^{-3})$	$1(10^{-4})$	$1(10^{-5})$	$1(10^{-6})$	$1(10^{-7})$
β	1.28	2.32	3.15	3.77	4.0	4.5	5.0

If $\boldsymbol{\beta} > 5$: $P_f = \boldsymbol{\Phi}^*(-\boldsymbol{\beta}) = [1/\sqrt{2\boldsymbol{\pi}}]\{(\boldsymbol{\beta}^2 - 1)/\boldsymbol{\beta}^3\}\exp(-\boldsymbol{\beta}^2/2)$

$\boldsymbol{\beta} = (1 - 0.118)/0.13 = \mathbf{6.78} > 5$ and $P_f = \boldsymbol{\Phi}^*(-6.78) = 0.4\{(6.78^2 - 1)/6.78^3$ $[\exp(-6.78^2/2)] = 6.0(10^{-12}) << (10^{-8})$

Fatigue limit (number of cycles) $N_f = (pm)/2\pi = 0.08(100)/2\pi = 1.27$

Maximum temperature at fatigue failure: $T_{max} = \boldsymbol{\theta}_{max}(0.1T_*) + T_* = 9(0.1)$ $600 + 600 = 1140[^0K] = \approx 840\ [^0C]$. Consider now the case that the inverse probability $P_f = 10^{-8}$ is given, then:

$$P_f = [1/\sqrt{2\boldsymbol{\pi}}]\{(\boldsymbol{\beta}^2 - 1)/\boldsymbol{\beta}^3\}\exp(-\boldsymbol{\beta}^2/2) = 10^{-8} \quad (5.33)$$

Solving Equation(5.33) we have: $\boldsymbol{\beta} \approx \mathbf{5.7}$

$$\ln[1/\sqrt{2\pi}] + \ln\{(\boldsymbol{\beta}^2 - 1)/\boldsymbol{\beta}^3\} - (\boldsymbol{\beta}^2/2) = -18.4 \quad (5.34)$$

One can see from "forward" solution that $\beta >> 1$, therefore the Equation (5.34) can be approximately reduced to: $\ln\{1/\boldsymbol{\beta}\} - (\boldsymbol{\beta}^2/2) = -18.8$. Let's try $\beta = \mathbf{5.7} \rightarrow -18 \approx -18.8$. Now the final result is: maximum dimensionless creep-fatigue temperature $\boldsymbol{\theta}_f = 8 - \boldsymbol{\beta}\,\boldsymbol{\sigma}_\omega = 8 - 5.7(0.0507) = 7.71 < 8.0 \rightarrow N_f = 7.71(0.08)(10^2)/(2\boldsymbol{\pi}) = =9.82$. Maximum temperature: $T = 7.71(60) + 600 = 1062.6^0K \approx 762^0C$.

Similarly, analytical models can be computed in the cases of Weibull and lognormal distributions of the fatigue limit, which allow predicting cyclic durability and constructing probability fatigue curves. The construction of such models based on a joint analysis of reference data on the characteristics of fatigue resistance and the results of full-scale fatigue tests of prototypes or their analogues makes it possible to carry out the resource design of critical elements of aerospace structures.

Conclusions

1. A phenomenological model for the quantitative assessment of cyclic creep and fatigue of composite and nanocomposite materials under conditions of high thermal stress states is proposed.
2. A mathematical model of the mechanics of damaged medium, which simultaneously takes into account the processes of viscoelastic deformation and damage accumulation in structural composites and nanocomposites with degradation mechanisms combining material fatigue and creep, is developed.
3. The material parameters and functions of the constitutive equations, which are necessary to describe the laws of cyclic viscoelastic deformation processes, are established.
4. The adequacy of the methodology for obtaining composite components parameters and scalar functions of the defining S – N curves for fatigue and creep is assessed.
5. The developed model allows predicting fatigue strength taking into account the influence of the orientation of the main directions of the stress tensor relative to the planes of elastic symmetry of the material.
6. The models use approaches based on modern continuum damage mechanics, and a technique for identifying the parameters based on the minimum necessary set of experimental data developed.

7. Criteria established for the maximum dimensionless temperature at fatigue failure stage of a composite and nanocomposite.
8. Verification of the analysis methodology of composites and nanocomposites under various conditions of high temperature loading is provided.
9. Deterministic approach of obtaining S – N curves analytically presented and many step-by-step examples are provided.
10. An applied probability approach is also provided (in the correlation theory of a probability framework). The input parameters and results are checked by using the "forward" and "inverse" probabilistic methods.
11. By the way, this revision was caused by the presence of two 5th chapters in one fifth chapter (!?). So this issue is now resolved.
12. In this book (as noted earlier), two creep-fatigue criteria are used—the fatigue of composites and nanocomposites, namely: the classical method (developing the S – N fatigue curve) and another criterion associated with finding the Damage Function, f(**ω**), and widely known damage criterion, f(**ω**) = 1. In this case, the first method is recommended to be used when the oscillation amplitude, a_0, is an independent variable, and the second method is recommended to be used when the damage function f(ω) = 1, that is, the entire cross section of a composite or nanocomposite is damaged.

References

[1] Van Paepegem W. and Degrieck J. Fatigue Damage Modeling of Fiber reinforced Composite Materials. Review, Applied Mechanics Reviews, 54(4): 279–300, 2000.

[2] Razdolsky L. Fatigue-Creep Phenomenological Models of Composites and Nanocomposites. AIAA Propulsion and Energy Forum, 2019.

[3] Harris B. Fatigue in Composites. CRC Press, Boca Raton, FL, 2003.

[4] Hashin Z. and Rotem A. A fatigue criterion for fiber reinforced composite materials, Journal of Composite Materials, № 7: 448–464, 1973.

[5] Hashin Z. Cumulative damage theory for composite materials: residual life and residual strength methods. Composite Science and Technology, № 23: 1–19, 1985.

[6] Lemaitre J. ed. Handbook of Materials Behavior Models. San Diego, Academic Press, 2001.

[7] Lemaitre J. A Course on Damage Mechanics. Berlin: Springer,1996.

[8] Kattan P. I. and Voyiadjis G. Z. Damage Mechanics with Finite Elements. Berlin, Springer, 2001.

[9] Talreja R. Fatigue life modeling. Second International Conference on Fatigue of Composites, Williamsburg, VA, 2000.

[10] Degrieck J. and Van Paepegem, W. Fatigue Damage Modeling of Fiber-Reinforced Composite Materials: Review. Applied Mechanics Reviews, 54(4): 279–300, 2001.

[11] Talreja R. Stiffness properties of composite laminates with matrix cracking and interior delamination. Engineering Fracture Mechanics, 25(5/6): 751–762, 1986.

[12] Hayder Al-Shukri and Muhannad Khelifa. Fatigue study of E-glass fiber reinforced polyester composite under fully reversed loading and spectrum loading"Engineering & Technology, 26(10), 2008.
[13] Hader Al-Shukri. Experimental and theoretical investigation into some mechanical properties of E-glass polyester composite under static and dynamic loads. Engineering and Technology, 2007.
[14] Rita R. and Bose N. R. Behavior of E-glass fiber reinforced vinyl ester resin composites under fatigue condition. Bulletin Materials Science, 24(2): 137–142, 2001.
[15] Razdolsky L. Phenomenological Creep Models of Composites and Nanomaterials Deterministic and Probabilistic Approach CRC Press (Taylor & Francis Group), Boca Raton, FL, 2019.
[16] Philippidis T. P. and Vassilopoulos A. P. Fatigue design allowables of GRP laminates based on stiffness degradation measurements. Composite Science and Technology, pp. 2819–2828, 2000.
[17] Razdolsky L. Probability Based High Temperature Creep of Composites and Nanocomposites. AIAA Space X, (AIAA Propulsion and Energy Forum), 2017.
[18] Philippidis T. P. and Vassilopoulos A. P. Complex stress state effect on fatigue life of GRP laminates, experimental. International Journal of Fatigue, № 24: 813–823, 2002.
[19] Muliana A. 2013. "Nonlinear viscoelastic – degradation model for polymeric based materials. International Journal of Solids and Structures, Elsevier Publishing Co.
[20] Mosegaard K. and Tarantola A. Probabilistic Approach to Inverse Problems. International Handbook of Earthquake & Engineering Seismology (Part A), Academic Press, pp. 237–265, 2002.
[21] NASA/TM—2010-216104, Report. Probabilistic Simulation for Combined Cycle Fatigue in Composites. NASA, 2010.
[22] Razdolsky L. Probability- Based Structural Fire Load. Cambridge University Press, UK, 2014.
[23] Lardner, R. Theory of random fatigue. Journal of the Mechanics and Physics of Solids, 15(3): 205–221, 1967.
[24] Lemaitre J. Course on damage mechanics. Springer Science & Business Media, 2012.
[25] Lemaitre J. and Desmorat R. Engineering damage mechanics: ductile, creep, fatigue and brittle failures. Springer Science & Business Media, 2005.
[26] Lemaitre J., Sermage J. and Desmorat R. Two scale damage concept applied to fatigue. International Journal of Fracture, 97(1-4): 67–81, 1999.

Index

D

E

F

G

H

I

K

L

M

For Product Safety Concerns and Information please contact our EU
representative GPSR@taylorandfrancis.com
Taylor & Francis Verlag GmbH, Kaufingerstraße 24, 80331 München, Germany

www.ingramcontent.com/pod-product-compliance
Lightning Source LLC
Chambersburg PA
CBHW060406220326
41598CB00023B/3032
9781032213026